# Barns of Wisconsin

→ Places along the Way ←

Richly illustrated with historic and contemporary photos, the Places along the Way series links Wisconsin's past with its present, exploring the state's history through its architecture.

**Fill 'er Up: The Glory Days of Wisconsin Gas Stations**
Jim Draeger and Mark Speltz
photographs by Mark Fay

**Encore: The Renaissance of Wisconsin Opera Houses**
Brian Leahy Doyle
photographs by Mark Fay

# BARNS

## *of* Wisconsin

# Jerry Apps

## photographs by Steve Apps

### foreword by Richard (Dick) Cates Jr.

### Wisconsin Historical Society Press

Published by the Wisconsin Historical Society Press
*Publishers since 1855*

Text © 2010 by Jerold W. Apps

First edition published 1977. Second edition published 1995.
Wisconsin Historical Society Press edition 2010
14 13 12 11 10    1 2 3 4 5

Publication of this book was made possible in part by grants from the Amy Louise Hunter fellowship fund and the D. C. Everest fellowship fund.

## wisconsinhistory.org

Printed in Wisconsin, U.S.A.
Designed by Brad Norr Design

Library of Congress Cataloging-in-Publication Data
Apps, Jerold W., 1934-
  Barns of Wisconsin / Jerry Apps ; with photographs by Steve Apps.—3rd ed.
    p. cm.
  Includes bibliographical references and index.
  ISBN 978-0-87020-453-1 (hardcover : alk. paper)  1.  Barns—Wisconsin—History.
  I. Apps, Steve. II. Title.
  TH4930.A66 2010
  728'.92209775—dc22

                 2009049285

Page iii: Early barns took many shapes; an L-shaped barn was not uncommon. Washington County, Highways 33 and 175. Page vi: This gambrel-roof barn has lightning rods and a metal roof ventilator. Pierce County, Highway 35 and 1150th Street, Prescott. Page viii: Quarried block silo. Oneida County, off River Road near the fairgrounds. Page xii: Interior view of the hay storage area in a Sears barn. Page 190: This quarried-rock barn was built in 1861 as a forty-four-stanchion dairy barn. It was restored in 2002 and is on the National Register of Historic Places. The farm where the barn is located has been in the Blum family since 1848. Green County, W6303 Hefty Road, Monticello. Page 192: Farmers sometime adorn their barns and other farmstead buildings with murals depicting favorite topics. Dane County, Hoepker Road north of Madison. Page 196: Nelson Dewey, Wisconsin's first governor, once owned an elaborate, 2,000-acre agricultural estate near Cassville. The barn, built circa 1868, is now part of the Wisconsin Historical Society's Stonefield historic site near Cassville. Page 200: The barn provided a constant backdrop to the farm family's life. WHi Image ID 26297.

To all those people who are working to preserve their barns and farmsteads
for tomorrow's generation

# Contents

# Foreword

**J**erry Apps's classic *Barns of Wisconsin* is a gift to every citizen of the state and the agricultural Midwest. Beyond nostalgia for our rural heritage, beyond the fascinating and ingenious examples of folk architecture, the barns that Jerry has immortalized in these pages stand as a lasting testament to our unique struggles and triumphs as agrarians from distant lands who came together in this welcoming wilderness and made a life, a culture, and a society. Europe may have its cathedrals, but Wisconsin's story is warmly told through its glorious barns.

These structures speak volumes about a time when most of us grew our own food, were intimate with the land we lived on, and shared our days with the animals we depended on for labor, transportation, clothing, and sustenance. Each barn instructs us about self-sufficiency, inventiveness, problem-solving, thrift, integrity, and other values we hold dear.

Perhaps the most poignant fact about these edifices is that we don't build them anymore. Our lives, our technology, our culture have moved on. People who own an old barn know they have something special, something that won't come our way again.

The cornerstone of the old barn on our family farm in Iowa County displays 1893 as its completion date on what was then the Thomas Stapleton farm. Its two-foot-thick foundation of native quarried sandstone and limestone is built into a hillside, which, as Jerry Apps teaches us, makes it a bank barn. Stapleton family lore tells us that the mason earned a dollar a day for the thirty days it took him to lay up the foundation. Zuber Hanson, a Norwegian craftsman, and his helpers constructed the 32-by-56-foot and 40-foot-high-at-the-cupola post-and-beam, wooden-pegged frame using white pine logs that had been floated down the Wisconsin River, then milled in Spring Green at the King Lumber Company: timbers, boards, and labor, $1,500. The Irish, German, and Welsh community of Jones Valley raised the barn together, just as they likely did for the barns of the contiguous neighbors, Carmody, Kraemer, Ryan, and Lloyd-Jones.

Though I farm here, we have long ceased using our old barn for the purposes it was intended for. But the vivid images of my parents and sister packing the mow elevator with small square bales of hay, my three brothers and I in the mow stacking thousands of bales a summer—bare-backed, sweat-glistened, dusty boys striving to be men—are a powerful part of my heritage.

Out of kindness, or perhaps in celebration of an exalted symbol, my family and I chose to renew our barn. Its restoration stretched over decades: rebuilding portions of the stone foundation, mow floor, and walls; block-and-tackles fastened to trees on the hillside, little by little pulling the structure back up straight from one hundred years against southwest winds; three layers of shingles-off, new-shingles-on; rebracing; a hundred-year barn party in 1993 (and many more parties since then).

For me, the old barns are a source of wonder and awe and are important to preserve in their own right. These barns, built by practical folks of modest means with hand tools, were not built simply to be functional; they were constructed with such care as to span generations, testament to hope and love of place. These builders valued commitment and community, and they shared an optimism that the fruits of the good earth, well tended, would serve their children, and their children's children, and beyond. In a culture where everything else seems to change so quickly, these barns are evidence of the enduring faith of our forebears.

So, as I have also grown older, the barns have taken on another, deeper meaning for me: they are a symbolic tie that binds the next generations to all those who came before, tending the land and livestock.

For the past decade and a half I have been a teacher at the University of Wisconsin in Madison, home to a magnificent dairy barn, still in its glory, constructed in 1897. As author Apps reminds us, this is the only barn in America designated as a National Historical Monument. Fittingly, it is on this campus that I have the privilege to assist the next generation of would-be farmers get a start at fulfilling their dream to farm. In this role it is my responsibility to teach to these young people the skills they need to be successful—and just as significantly, to help them discover and cherish what is most valuable about the culture of agriculture for which they are to become the torchbearers.

Jerry Apps and his son Steve, who is credited with the splendid photographs in this edition, traveled the countryside of Wisconsin and found that many of the barns that graced the pages of earlier editions of this book are no longer standing. The stone foundations with tie-stalls or stanchions for milk cows are being replaced with useful structures less expensive to build: well-ventilated hoop buildings and free-stall barns, and parlors for milking that demand less labor and save the farmers' knees to boot. Farming is changing, moving on, as well.

But Jerry Apps, not one to let our barns disappear unheralded, entreats us to identify and cherish "what of the old is worth keeping." I am with him: the old barns of Wisconsin are our legacy; our collective treasures. They should be cherished, and preserved, lest the values of the culture that bequeathed these gifts pass on with them.

—*Richard (Dick) Cates Jr.*
*Spring Green, Wisconsin*

*Dick Cates is senior lecturer in the Department of Soil Science and director of the Wisconsin School for Beginning Dairy and Livestock Farmers, a program of the Center for Integrated Agricultural Systems and Farm and Industry Short Course in the College of Agricultural and Life Sciences at the University of Wisconsin–Madison. He serves on the Board of Directors of the Wisconsin Department of Agriculture, Trade and Consumer Protection. Dick and his wife, Kim, and family own and operate Cates Family Farm LLC, a grass-fed beef business in Wyoming Township, Iowa County. Dick is the author of* Voices from the Heart of the Land: Rural Stories that Inspire Community.

# Preface

It was 1975, and I was a professor in the College of Agricultural and Life Sciences at the University of Wisconsin. The phone in my office on the Madison campus rang. The caller was Glen Pound, then dean of the college.

"Could you stop down by my office this afternoon?" the dean asked. In those days I tried to avoid deans, but this one was my boss, and I soon was sitting across from him in his big office, where a huge John Steuart Curry painting hung on one wall.

I expected the worst. Had someone complained about me to the dean? Did I mess up my expense account, again? (I had a joint appointment with UW–Extension and did a fair amount of traveling around the state.)

"A friend of mine has several sketches and paintings of barns," Dean Pound began.

"Yes," I said, not knowing where the conversation was going.

"I was wondering if you'd have a little time to meet with him. His name is Allen Strang, and he's an architect here in Madison." Dean Pound explained that his friend wanted someone to write captions for his barn pictures. He figured I could do it in a weekend.

"Sure," I said. I had learned long before then to never say no to a dean.

I met with Allen a week later for lunch. We had a delightful conversation. I quickly learned that Allen, a city boy, knew little about barns, except he found them great subjects for sketching and painting. He wanted to publish the pictures, but they needed captions. Would I be willing to write them?

"I'll try," I said. Little did I know that those few "weekend captions" would lead to more than a year's worth of writing and research, with the book *Barns of Wisconsin* the result. I had grown up on a small dairy farm in central Wisconsin and had spent many years in a barn—milking cows, forking back hay in the haymow, pitching manure, tossing down silage, the sorts of things farmers do without giving it much thought. With my personal experience, plus extensive research, I had plenty of material. But would anyone be interested enough to buy an entire book on the topic?

*Barns of Wisconsin* came out in 1977. It quickly became one of my most popular books, winning several awards and selling many thousand copies. More than thirty years later, people's interest in barns continues, perhaps stronger than ever. For this revised edition, published by the Wisconsin Historical Society Press as part of their Places along the Way series, I collaborated with my son, Steve, who is a photographer for the *Wisconsin State Journal* and has worked with me on several book projects. Steve has traveled many miles shooting barn pictures throughout the state.

This new edition also contains detailed information about a generation of barns constructed after the post-and-beam era, built with trusses made of lighter materials than the much heavier posts and beams. Also updated in this volume is a list of notable Wisconsin barns on the National Register of Historic Places, with locations. The captions for contemporary barn photographs also include locations, and all these barns stood tall at the time of publication. But Wisconsin barns are disappearing at an alarming rate, and some of these barns may no longer exist by the time you read this. Also note that almost all of these barns are on private property. Please respect the privacy of the owners as you travel the state on barn-hunting tours.

Many people have assisted me on this project. They include Charles Law, University of Wisconsin–Extension's Barn Preservation Program; Ruth Olson, associate director for the Study of Upper Midwestern Cultures, University of Wisconsin–Madison; Marty Perkins, Old World Wisconsin, Eagle; Allen Schroeder, Stonefield historic site, Cassville; the staff of Circus World Museum, Baraboo; Roger Springman, former president of Barns Network of Wisconsin; Doug Cieslak, Driftless Area Land Conservancy, Dodgeville (Thomas barn); John (Bud) Carroll, Green Lake Conference Center, Green Lake; Howard Sherpe, author, Westby; the staff of the Vernon County Historical Society; Jim Draeger, Daina Penkiunas, and the rest of the staff of the Wisconsin State Historic Preservation Office; and many more who gave me tips about barns, showed me photos of their barns, and told me barn stories.

As always, a special thank-you to my editor, Kate Thompson at the Wisconsin Historical Society Press, and to my wife, Ruth, who tirelessly reads all of my manuscripts.

Arched-roof barns became popular after World War II. Here, a smaller arched-roof addition has been attached to the larger barn. Note the poured-concrete silo. Clark County, Highway 29 and Sterling Avenue east of Thorp.

In 1966 this was what remained of the farmstead on our farm. The old barn was falling down; the house had burned in 1959. Our daughter, Susan, is in the foreground. From the author's collection

# Introduction

Our neighbors began describing the old barn as an eyesore. The barn wasn't old—for a barn. It was built in 1910. But it looked old, like a man who is wrinkled and bent and clearly in the autumn of his life. Each year the barn leaned a bit more to the south, toward a rather sturdy combination granary, chicken house, and wagon shed that stood a few feet from it. The barn boards had never been painted, and the wood had weathered to several shades of gray. I could feel the wood grains standing out when I ran a finger slowly across the boards.

My wife and I discussed the old barn at some length shortly after we'd acquired the central-Wisconsin property where it stood. We talked about straightening and repairing it, hoping to preserve its basic beauty. But everyone we consulted said it was no use even to try; the building was too far gone. With considerable reluctance and much sadness we finally decided the barn had to come down.

Most of the summer we worked at tearing down that old barn. We tried to salvage as many of the boards and timbers as we could. In many ways it was a sad task, for a barn seems to have a life of its own, and we were destroying this one. During the salvage operation we learned that our old barn was an exception, that the vast majority of the barns built a decade or two before ours and a decade or two after still stood straight and true and continued to be used.

Our barn had not been built well. Its framing materials—primarily two-by-fours, two-by-sixes, and a few two-by-eights—had been too light, and over the years the windstorms had taken their toll. With each year's storms, the barn leaned a bit more. The farmer who'd owned the place before we acquired it had tried to brace it with cedar fence posts. But the posts hadn't helped. During the last years he owned the barn, the farmer was afraid to keep his livestock in it, so the barn stood empty. It wasn't completely empty, though, for pigeons continued to roost in the haymow, field mice found homes in an abandoned straw pile in one corner, and barn swallows lived comfortably in their mud-and-straw nests tucked up against the ceiling of the stable.

By the end of the summer, the barn was down. All that remained was the concrete foundation—it was as sturdy as the day it was poured—and a pile of boards and framing materials. During the process of demolishing that old barn I came to respect its builders. Though they had made some serious errors, the barn was nevertheless no small task to raze.

I've known barns for a long time. Born and raised on a central-Wisconsin dairy farm, I spent many hours of my growing-up years in our barn. In winter, after weeks of cold and snowy weather, I looked forward to the days when we could turn the cattle out into the barnyard to romp in the snow like yearlings. How fresh and clean the stable smelled when we once more brought them into the barn and locked their necks into the stanchions. We had carried forkfuls of fresh straw from the dwindling strawstack out back of the barn and had bedded their stalls. And we'd piled alfalfa hay high in front of them, from the haymows in the upper barn.

During the most intense cold, frost had accumulated on the barn windows until I could scrape it off by the handful. Warm days in late winter encouraged the frost on the windows to melt and drip. The sun poured through the windows, reflecting the dust particles from the hay. A few weeks later, when the snow had melted and the green grass appeared on the south side of the pasture hills, the cows were turned out during the day and confined to the stable only during the cold spring nights.

In winter the old dairy barns were warm and cozy, with thick frost on the stable windows and corn silage, alfalfa hay, oat straw, and cow manure smells combining to create a pleasant, earthy experience. Dodge County, County Highway P south of State Road 60.

In these early spring days, when we were busy with the planting, we worried about wind-storms, which are most prevalent in the Midwest in spring. Unfortunately, this was also the time when barns were most vulnerable, for the haymows were empty. On a windy day in spring I'd crawl up to the peak of our barn, look out over the countryside, and see clouds of dust rolling across the freshly tilled fields. I could feel the wind shake the barn and hear the oak timbers protest loudly as they squeaked and snapped with each gust. My father knew, and I knew, too, that a tornado could easily destroy our barn, tear off its roof, and twist its timbers into a pile of useless rubble. I also knew a strong straight wind, the kind we often got in the spring, could push our barn off its foundation and create havoc. I knew this because it had happened one day in May, collapsing our barn wall and killing several head of livestock.

By late June we started filling the barn with hay. The McCormick mower chattered around the hayfield, slicing off the stems of clover, alfalfa, and timothy. After the hay had a day or so to dry, my father raked it with our high-wheeled dump rake, and my brothers and I piled the hay into bunches to dry even more. When the hay was just right, and the weather sunny and warm, we hitched the team to the hay wagon and started hauling hay to the barn. Slowly the steel-wheeled wagon creaked around the field as we piled the hay higher and higher until we couldn't fork on another bunch. When we arrived at the barn, my father slowly steered the team and wagon up the incline onto the barn's threshing floor. When the wagon was directly under the hayfork, he yelled, "Whoa!" Meanwhile, I hitched a third horse to the rope.

My father grabbed the slender trip rope and began to pull. The hayfork moved along its track in the peak of the barn, tripped when it arrived directly above the wagon, and settled onto the load of hay. My father took the heavy, two-tined fork—it was about three feet long—and thrust it deep into the hay before setting the trip mechanism.

"Take 'er up," he called when everything was ready.

I clucked to Dick, the black horse hitched to the hayfork rope, and the rope tightened. Slowly the forkful of hay lifted from the wagon as the hayfork rope moved through the series of pulleys that connected the horse to the hayfork thrust into the hay. The wooden pulleys creaked and protested as the mound of hay lifted from the wagon and rose to the hayfork track fastened just under the barn roof.

I heard a clunk when the hayfork hit the track and slid along it over the mow. When the hay was midway over the mow, my father pulled the trip rope, and a cloud of dust and leaves filled the air as the hay dropped with a whoosh. "Whoa!" I yelled to Dick when the rope went slack. I turned the horse around and returned to the barn to pull up another load. Meanwhile, my father climbed into the mow and forked the fresh hay into its far corners with a three-tined pitchfork. Sweat streamed from his forehead and soaked through his shirt. It was not uncommon for the temperature in the haymow to rise considerably over one hundred degrees. Five or six hayfork loads later, we returned to the field to gather another wagonload of hay. This procedure continued, day after day, until the hay fields were bare and the mows were piled full to the hayfork track.

Often on a rainy day in summer, after the morning chores were finished, my father, my brothers, and I crawled up the wooden ladder to the haymow and sprawled on the fresh hay to talk and rest and listen to the raindrops patter on the wood-shingled roof. In dry weather I could see sunlight shining through the shingles in many places, and I always wondered why the roof didn't leak. But once the rain started, the cedar shingles swelled and filled the cracks.

Soon after the barn was filled with hay, the oat crop was cut, shocked, and threshed. The oat bins in the granary were filled—feed for the cows in winter. A huge strawstack once more stood behind the barn—bedding for the many months when the cows had to remain inside.

In September, if the rains had come at the right time and the sun had shone long and hot, part of the corn crop was ready to put in the silo—more winter feed for the cows. The silo, constructed of redwood staves standing vertically on a fieldstone wall, stood a few feet east of the barn. The silo was connected to the barn by a silo room, a small structure large enough to hold silage to feed the dairy herd twice, morning and evening. Once the cows came into the barn for the winter, silage had to be forked down from the silo once each day until the cows could be turned out to pasture the following spring or until the silage had been consumed. Too often the silo was empty several weeks before the pasture grass had grown tall enough for grazing. Then the cows had only hay to eat.

Usually one day in October the dark gray storm clouds rolled in from the west, a drizzly rain commenced to fall, and the cows were kept in overnight. Forkfuls of fresh straw were carried into the barn, and the cows were locked into their stanchions. When we went out to the barn for milking that evening, we were greeted with a strange combination of smells: wet cow, fresh straw, and warm, foamy milk. The cold autumn rain splashed against the barn windows and ran down in little rivulets. But it was comfortable in the barn. A kerosene lantern hung on a nail above the milk cans; another lantern hung behind the cows at the far end of the barn. Their flames cast a peaceful light. I could think of no more comfortable place to be on such a night.

Soon the weather became cold enough so the cows were kept in the barn both day and night and were let out every few days to exercise. By late November, the temperature often dropped below zero, and the cows were not allowed outside even to exercise.

When I climbed into the upper part of the barn to throw down hay, I was greeted with long strands of frost attached to the cobwebs. Some of the strands were more than six feet long, beautiful in the soft yellow glow of the kerosene lantern. Above the hay chutes—the openings from the threshing floor to the stable below—the hay was white with frost all the way to the roof of the barn. The hay chutes were used as ventilators, allowing the moisture-laden air from the stable to escape to the upper part of the barn. When the warm, moist air collided with the cold air, the moisture immediately turned to frost. After a few days of below-zero weather, frost hung everywhere in the upper parts of the barn, creating a scene of beauty. Pitching hay from the mows became a dangerous job, however, for the frost made the wooden ladders slippery.

During winter days, my father and I spent an hour and a half milking cows both before breakfast and after supper. Of course, the milking time was the same during the other seasons of the year as well, but in winter, additional time was required to carry feed to the front of the cows and to haul manure out from behind them. Like milking, these were daily tasks.

In recent years my interest in rural buildings, and particularly barns, has sharpened. Thousands of them have disappeared—torn down, bulldozed over, burned—in the name of progress, or what is called progress. These old barns and the land they stood on made way for shopping malls, condo developments, and paved parking lots—the relentless march of city into country.

**As cities continue their march into the countryside, we see the old contrasting with the new. Here a modern wind turbine dwarfs an old gambrel-roof barn. Fond du Lac County, intersection of County Highways Y and F.**

Many of these old barns simply stand alone, abandoned, visual memories of a changed agriculture. When I see an old barn rotting away, I often wonder if it might be the last barn of its type, a symbol of the past that can never be recovered.

Many old barns, on the other hand, are still filled with hay each summer and house dairy cattle as they did when first built. Many are approaching one hundred years of age yet are as sturdy as when first put up. Why have they managed to survive so well, particularly during an age when we expect things to wear out within a few years?

This question and more prompted me to learn as much as I could about barns, by observing them, by reading about them, and by talking to people who have intimate knowledge of these remarkable rural buildings. This book is what I found out. It is not a detailed account of how barns are built, though some information along these lines is included. Likewise, it is not a history of rural Wisconsin, although a good deal of history is woven through these pages.

This book ranges from the earliest barns—the barns the pioneers built when they first arrived in Wisconsin—through the era of post-and-beam construction and the lightweight construction movement that began in the early 1900s to, finally, a brief mention of present-day large, open-air free-stall barns that house hundreds, sometimes thousands, of dairy cows.

This book focuses most on old barns because I believe understanding them can help us understand something about the people who built them and spent much of their lives working in them. We can learn something about the values and satisfactions, the motivations and frustrations of these Wisconsin farmers of an earlier day. And in a broader sense, we can learn something about ourselves, for nearly all of us can trace our beginnings back to the soil.

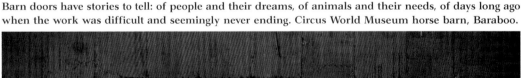

**Barn doors have stories to tell: of people and their dreams, of animals and their needs, of days long ago when the work was difficult and seemingly never ending. Circus World Museum horse barn, Baraboo.**

In a circa 1874 photo, a farmer shows off his flock of sheep and cattle with his big barn in the background.
WHi Image ID 26136

As the moon rises on a cloudless night, a barn and silo stand alone, a piece of faded history, testament to early Wisconsin farmers who saw a future in dairy farming. Dane County, Highway M near Governor Nelson State Park.

# 1

# BORN OF THE LAND

## *Barns and People*

Barns are many things to people. To travelers driving along highways, barns help make the rural landscape interesting and attractive. A red barn contrasted with fresh snow is a sight to behold. A white barn standing among a collection of farm buildings against the greens of summer presents a pleasing view to even the least aesthetically inclined. No matter what its color—red, white, black, green, or the gray of aging, unpainted wood—a barn evokes a feeling of beauty in the hearts of many.

To the romantic, barns are rich sources of lore. For thousands of people, barns trigger nostalgia, taking them back to a time when they believe living was simpler and life richer. To the architect, barns underline the theory that form follows function, that the shape and size of a barn relate to its intended use.

But for the farmer, a barn is central to the farming operation—pleasing to look at, to be sure, but more important, of economic value. A farmer regards the barn the way an industrialist regards the factory. Barns house the farmer's means of production. To the farmer, a barn must be functional first, no matter what else it may mean.

Even the word *barn* reflects the building's function. It comes from the Old English *bere,* meaning "barley," and *ern,* meaning "place" or "closet"—literally, a barley place or barley closet. Barley was an important crop in early times in Europe, and *barn* came to mean a place where barley was stored. The early definition of *barn* said nothing about shelter for livestock. The Old English word *byre* meant "cow house." The byre was often a separate building from the barn. Later the byre became a part of the barn, so a barn was thought of as both housing livestock and storing grain and hay.

Even in this country, during the colonial period, barns were used primarily for the storage and threshing of grain. The farmer piled the grain on the threshing floor and beat it with a flail to separate the kernels from the straw. The early barn was built around the threshing floor, which has remained to this day, although it is now more often referred to as the driveway, or barn floor.

# Barns and Rural Values

No one can deny that the United States was built on a rural foundation. The 1790 census reported only 202,000 people living in urban areas, compared with almost 4,000,000 living in rural areas.[1] Of course, today the reverse is true. The vast majority of the U.S. population lives in urban centers. The national trend followed in Wisconsin. In 1935 Wisconsin had the largest number of farms: 199,877. In 2005 the number of Wisconsin farms had declined to 76,500, only 38.3 percent of the peak number.[2] Likewise, the number of Wisconsin dairy farms has fallen. The state had 143,000 dairy farms in 1950;[3] by 2009 Wisconsin's number of dairy herds had fallen to 13,009.[4]

The barns sprinkled liberally throughout the countryside remind us of the hundreds of thousands of farms that once operated in Wisconsin and the values rural people held when these barns were built.

### Optimism

Though there were often hardship, disappointment, and much sadness on the farm, the barns we see today are evidence of a basic optimism about life on the land. But there's a contradiction here, too. Many farmers could scarcely make a go of it, finding it difficult and often impossible to pay their taxes and the interest payments on their mortgages. It was not at all unusual for a farmer to buy a farm only to lose it—through a series of bad years caused by poor weather, declining markets, failing health, or some other reason—and have to move on. Some farm families gave up and moved to cities and villages to earn a living. Many others moved onto other farms with the hope that things would be better. Sometimes they were right, but often they were not, and the cycle would repeat.

Nevertheless, the barns farmers built were symbols of optimism, for they were built not only for the farmer's expected use but for the farmer's children and grandchildren as well. Many of the big barns still seen throughout the Wisconsin countryside were constructed in the late 1800s and early 1900s. Though added to and often modified several times over the years, many of these barns continue to be used today—albeit often for purposes different from their original use. These barns were built to endure.

### Pride in Workmanship

Early Wisconsin farmers took great pride in their work, whether it was plowing a straight furrow, stacking straw neatly, or building a barn. And they expected those who worked for them to take pride in their work, too. Indeed, they insisted on it. The barn builders—most Wisconsin

**Early farmers were proud of their barns, their horses, and their families. Family photos were often taken in front of the barn.**
WHi Image ID 26442

farmers hired professional or semiprofessional help to build their barns—worked on the structure as if it were their very own. No detail was overlooked, no error camouflaged. A barn was constructed as well as the barn builder was able. When he finished his work, the builder was as proud of the barn as was the farmer who paid the bill. The barn builder would bring his friends to show them his work, and of course he'd bring other farmers who were considering building a new barn. A completed barn was the best advertisement for the talents of the barn builder. It was a part of him, just as a well-written book is a part of the author. Something of the builder became a part of each barn he built, and he would have it no other way. He knew no other way, for pride in his work was a guiding value in his craft. Of course the farmer worked with and assisted the barn builder throughout the process. Thus farmer became one with the barn, from the digging of the foundation to the nailing of the last board.

## Value of Time

Early farmers valued time as an investment. Today the emphasis is on efficiency, on doing things as quickly as possible, on doing more in less time, whether it's harvesting a field of oats or traveling to Chicago. Speed is of the essence.

Farmers were not opposed to speed, but many times it was necessary to be deliberate, to take more time to do a job. Putting together the timbers of a barn was one of those times. It might take several weeks for the barn builder and the farmer to select the timbers, saw them to size, cut mortises and tenons, and drill holes with a hand drill for the oak pegs that would hold the timbers together. Though it took a long time to build a barn using this method, the farmer and the barn builder knew a quality job took time. Time was an investment, not an enemy to be confronted at every turn.

## Love of Beauty

Whether they'd admit it aloud or not, early farmers prized beauty. Barns were constructed to be functional. But function and beauty seemed to merge in many barns. What was most functional as a building also had great beauty. Because most farmers valued beauty, they not only built functional, attractive barns but also kept them well painted and in good repair. The farmer wanted neighbors and those traveling by the farm to enjoy and admire the beauty of the farm buildings.

## Practicality

In the early days as now, farmers concerned themselves with the practical. Barns were built with a purpose in mind; only as a second priority were they something to show off. Occasionally a farmer had a fancy cupola constructed on the barn roof, purchased a weather vane a bit on the ornate side, ordered lightning rods adorned with glass bulbs of various colors. But the cupola was a part of the barn's ventilation system first and a decoration only second. The weather vane was primarily

Early barns were often adorned with decorative cupolas and weather vanes. The cupola also served as a ventilator, and the weather vane told wind direction. But important functions aside, they added an additional dash of beauty to the barn. **WHi Image ID 26135**

a guide for the farmer's weather predictions. Even the multicolored glass bulbs on the lightning rods served a practical purpose: a broken bulb suggested the rod had been struck by lightning, and the farmer should check to see that the ground wire was still functional—or watch to see if the neighbor boy was practicing with his BB gun, for the glass bulbs on lightning rods were enticing targets.

A few farmers, particularly those who were well-to-do and might rightly be called gentlemen farmers, may have built their barns to impress as well as to function. These barns were clearly exceptions. Ordinary farmers couldn't afford to impress their neighbors.

## Cooperation

Though the farmer is often thought of as a rugged individualist, the barn is a symbol of cooperation. Though the professional barn builder planned the barn and assembled the structural pieces, the farmer's friends and neighbors helped erect it. People gathered, often from miles around, on the day of the barn raising. Many hands working together were required to raise the massive timbers of a new barn into place. Only by the most careful cooperation was a barn raising possible, for the entire crew of neighbors and friends, under the direction of the barn builder, needed to work as a unit—pushing, lifting, pulling until the skeleton of the new barn was in place.

Farmers valued their neighbors and depended on them in many ways. Barn building was but one evidence of neighborly cooperation. Threshing, silo filling, corn shredding, and wood sawing were other examples of how farmers worked together throughout the year to help each other complete the tasks that the individual farmer couldn't do alone or could do only with great difficulty. Cooperation was an important value held dearly by the farmer; the farmer going it alone was a myth.

## Work Ethic

Without question, early farmers valued hard work. To the farmer, accomplishing anything worthwhile meant working for it. If a barn was to be functional, strong, and able to last for generations, then it obviously required time and effort to build. There was also a positive correlation between hard work and pride in what was accomplished. The more arduous the task, such as building a barn, the greater the reward and the more pride the farmer had when the job was completed.

## Barn Valued over House

Because farmers were extremely practical, they valued the barn above all other buildings on the farm, including the house. Money for improvements on the farm went into the barn, not into the house. "It's the barn that gives us a living," a farmer might argue. "The house is only a place to live."

On many farms, the barn was kept relatively up to date, though the house included only the most basic elements for living. It was not uncommon for farmers to devise a running-water system for the barn so the cattle could drink from automatic watering cups, while drinking water was carried to the house in a pail from the pump house. It was often years after running water had been installed in the barn that the pipes were laid to the house.

## Love of the Land

Of course the underlying value that all farmers held was (and is) love for the soil. Farmers had faith in the land. They believed that with hard work and careful attention to details of farming, they could make a living for their families from the soil—most of the time anyway. They believed that the only place to raise children was in the country, away from the enticements of city life where people lived too close together and children seldom had enough work to do. On the farm there was never a shortage of work, including work for the children.

Farmers prided themselves on the acres of land and the buildings that were an important part of the farming enterprise. They couldn't understand how a person could be happy living

The barn is the centerpiece of the farmstead, considered by most farmers to be far more important than the farm home. Dane County, Lacy Road, Fitchburg.

in the city, in a house on a quarter acre of land or, worse yet, in an apartment building. A farmer needed room to stretch out his arms and breathe deeply the country air, to stand by his barn and look out over the fields that were his. It gave him a feeling of oneness with creation; it made him feel he was a cooperator with nature. The farm gave him roots. The farmer knew where he belonged, and he knew what he believed.

The great barns of rural Wisconsin are testimonials to the farmers' values. These buildings stand against the elements year after year and are seen by travelers from all walks of life. Barns are much more than buildings that shelter cattle, horses, and feed. They are symbols of farm life and farmers. They tell much about those farmers, from the late 1800s until

## ⁂ Barn Raisings ⁂

**In the early days of barn building,** community barn raisings helped get the job done. A barn raising was both a community social event and a service for the farmer. On the day of the barn raising, from sixty to a hundred or more people traveled to the building site from miles around. Announcement of the barn raising was passed

Chester Herschberger of rural Westby has a new barn, thanks to the joint efforts of Amish community members who took part in this barn raising on July 18, 2008. **Courtesy of Dorothy Jasperson, Westby Times editor**

by word of mouth and by scribbled notices at the feed mills, grocery stores, hardware stores, taverns, and barbershops throughout the area. By eight o'clock the first barn raisers started arriving, usually accompanied by their wives and children bearing food for the huge potluck dinner that was served to all at noon. Under the careful direction of the head carpenter, the men began lifting the bents into place. With a combination of pike poles—long poles with sharp spikes fastened at one end—ropes and pulleys (sometimes assisted by horses), and brute human muscle power, the huge timbers and braces were lifted into place. If the carpenter and crew had measured carefully, the bents slipped easily into the notches cut in the huge timber sill.

Only the skilled climbers were allowed to work on the upper parts of the barn. These usually included the carpenter's helpers and a few others from the community who had experience in barn raising and walking along the beams that stuck into the air as high as forty feet.

When the bents were in place and the connecting beams were erected and pinned, the roof rafters were attached. One man, likely the most skilled climber of the crew, worked his way along the ridge of the barn, fastening together the rafters as they were pushed up to him. At the same time, he attached the hayfork track to the ridge of the barn with lag screws. When the rafters were in place, the crew immediately began nailing the present time. They record the changes in agricultural technology. They are reminders of ethnic groups that settled various parts of the state. They tell of the agricultural history of Wisconsin, from the time farmers depended almost solely on wheat growing until now, when dairy farming, other types of livestock farming, and a host of alternative agriculture pursuits such as organic farming, commercial vegetable and fruit growing, and Community Supported Agriculture (CSA) farms have emerged.

The barns of Wisconsin are history books in red paint, sociology with gable roofs, theology with lightning rods. In many ways, barns are Wisconsin agriculture nailed together in buildings with cupolas on the top. Here is where both life and death on the farm often occurred simultaneously; here is where farm boys and girls learned about responsibility, where many a farmer began and ended his day 365 days a year for his entire life.

---

roof boards onto them. The boards were also nailed to the sides of the barn, time permitting, on the day of the barn raising. If a large crew had come, it was not unusual to have the framing of the barn in place, all of the roof boards nailed onto the rafters, and most of the siding completed.

Meanwhile, the women and girls were preparing a huge meal and visiting with one another, having come together in greater numbers than was true at the annual country-school picnics and Christmas programs. A barn raising had a festival air. There were foods of many kinds, including a wide variety of homemade pies and cakes. The children, kept at a safe distance from the workers, met new friends and enjoyed a day of play while watching the progress on the barn.

By chore time, when the farmers had to return to their farms for the evening feeding and milking, the farmer had a new barn standing in his yard. During the next few days the carpenter and his helpers completed nailing the siding into place and nailed the cedar shingles onto the roof boards. Sometimes extra men were hired to help with the shingling; as many as ten or a dozen men might work on the roof at one time. Using a shingling hatchet, a person could quickly nail the wood shingles into place. The shingles were pulled up onto the roof with the hayfork rope and pulleys that would become a part of the hayfork system when the barn was completed. But before they were pulled up to the roofers, each bale of shingles was doused with water to make the pieces less likely to split when they were nailed into place. The shinglers started at the eaves of the roof and worked toward the ridge, overlapping the shingles so that only four inches of each one was exposed to the weather. When the barn was finished, the carpenter and his crew left, leaving the farmer to decide if the barn should be painted or not.

When lightweight construction became more popular for barn building in the early 1900s, barn raisings continued, but because the big, heavy posts and beams were no longer a part of the construction, fewer workers were needed to erect the barns. Barn raisings continue today in Amish communities, following the age-old practice of neighbors helping neighbors to accomplish a major task.

Many German immigrant farmers settled in eastern and southern Wisconsin. They often used a half-timber building approach for their barns and other buildings. Between the timbers they filled in with nogging, in this case bricks. Waukesha County, Old World Wisconsin Visitor Center, Eagle.

# 2

# SETTLERS' STRUCTURES

## *Pioneer and Ethnic Barns*

Although Jean Nicolet, a Frenchman, pulled his boat ashore north of the site of Green Bay in 1634, many years passed before European agriculture and its associated buildings came to Wisconsin. It wasn't until 1819 that substantial numbers of people began arriving in Wisconsin, paddling up the Mississippi from Illinois and Missouri to the lead mines in the southwestern part of the state. To these settlers, mining offered economic security, and farming provided food for their families.

After Indian cessions in 1832 and 1833 and the government land surveys—first of the southwestern part of the state (1835) and then of the southeastern part (1836)—farmers began arriving in Wisconsin. These tillers of the soil were mostly Yankees, from New England, Pennsylvania, and primarily from New York. The Erie Canal had opened in 1825, providing a route for the New Yorkers, via the Great Lakes, to Wisconsin. The Yankees left behind depleted eastern soil, high land costs, and declining markets for their wheat. Wisconsin, with its cheap land and small population, beckoned as a place to start a new life.

When the newcomers arrived in the state, they quickly built log cabins for their families. Next they hurriedly erected shelters for their work animals, usually a pair of oxen. These first animal shelters were often crude lean-tos made of logs and tree branches and thatched with grass and bark. Yankee pioneers were not much concerned about sheltering the cow or two they owned. Among the pioneers' possessions, the milk cow was of minor importance. She found her keep in the woods and swamp meadows. The women milked her from the time she freshened in the spring until she dried off in the fall. During the winter, the cow humped her back and crowded behind a strawstack to escape the cutting winds and frigid temperatures.

Some Yankee pioneers constructed log barns that were more substantial than the early lean-tos, but the barn's purpose was to shelter the work animals, not the milk cows. Yankee farmers who settled in Wisconsin came intent on growing wheat, not livestock. They were most concerned, after building minimal shelter for their families and work animals, with clearing land and planting the first crop of wheat. These early Yankee pioneers settled primarily in Kenosha, Racine, Walworth, and Rock Counties.

Pioneers often built log barns when they arrived in Wisconsin. These barns provided a rustic shelter from Wisconsin's harsh winters, but little else. At this farmstead, the hay was stacked outside. Alma, Wisconsin, date unknown. **WHi Image ID 25066**

From 1836 to 1847, the years Wisconsin was a territory, the population grew from 11,683 to 210,546. (Wisconsin became the thirtieth state in 1848.) Settlers streamed into Wisconsin from Ohio, Indiana, Michigan, Virginia, Kentucky, Tennessee, and North Carolina, besides those states already mentioned.

Beginning in the 1830s and continuing into the 1900s, immigrants from Europe began selecting Wisconsin as their new home. The Germans were the first large group that arrived with the intention of farming. Up to about 1846, German immigrant farmers settled primarily in Washington, Ozaukee, Milwaukee, Jefferson, and Dodge Counties. Following the lead of the Yankee settlers who had preceded them, they built log cabins for their families, erected simple log shelters for their work animals, and began clearing land to grow wheat as quickly as possible.

The Germans, like most settlers, wanted land that combined some wooded area for building purposes, some high prairie because it drained well and was easy to break, and some low prairie or meadow for growing hay. But much of this preferred land, particularly in southeastern Wisconsin, had already been settled by the Yankees, who were often willing to sell their "developed" farms, but at prices considerably above the $1.25 per acre charged by the United States government for federal lands. So the Germans were forced to settle lands farther north. By 1850 German immigrants were wresting farms out of the forested wilderness of Sheboygan, Manitowoc, and Fond du Lac Counties.

In the early 1830s Norwegians, too, began arriving in Wisconsin, first in Rock County; then in Waukesha, Racine, and Dane Counties; and then westward to Vernon, Crawford, La Crosse, and Trempealeau Counties. During the 1840s and 1850s, other ethnic groups settled in Wisconsin—Irish, Welsh, Scots, Dutch, Swiss, Luxembourgers, Belgians, Bohemians, Icelanders, Danes, Swedes, and Italians.

The Charles Koehler family gathered for a photo in front of their rustic log barn. Black River Falls, 1890. WHi Image ID 1917

The Poles arrived in Wisconsin from the 1850s until the early 1900s. They settled primarily in central and northeastern Wisconsin, with a large concentration in the Stevens Point area. The Finns were relatively late to arrive in the state, beginning in the late 1800s and continuing into the early 1900s. They settled mostly in the far north, in Douglas and Bayfield Counties. By 1900 Wisconsin's population of two million included people from more than thirty countries. (In 1930 Germans were the largest nationality group in Wisconsin; Poles were second and Norwegians third.)

Log buildings were universally constructed by both the Yankees and the ethnic settlers in the state. These simple structures could be built from materials taken from the land when it was cleared. And the same tool, the lowly ax, could be used to clear the land and build homes and barns as well. The immigrants brought with them from the Old Country the skills necessary

On isolated Madeline Island in Lake Superior, a Swedish immigrant built a two-story log barn with a stable below and hay storage above. Ashland County, east of La Pointe on County H.

for building log structures. Solid-log and hewed-timber construction had a long history in Europe. This building approach, found primarily in the rural areas, was common as far north as the Scandinavian countries, south to Switzerland, west into Germany, and east into Russia and the Balkan countries.

The building of log structures marked the advance of the frontier in Wisconsin. Though the early settlers almost always built log cabins and log barns, they all, it seemed, strove to replace the log structures as soon as possible with buildings of wood-frame construction. Sometimes they covered the sides of their original log cabin with wood siding and attached wood-frame additions to it. They also did this occasionally with their log barns. In both instances the log structure remained but was hidden. When a wood-frame barn was built, the old log barn often continued to serve—as a hogpen, sheep barn, or shelter for the chickens.

By 1870 few log buildings were being erected in the southern part of the state. (Some buildings of mill-sawed logs, not hand-hewn logs, were built as late as the Depression years of the 1930s.) In the north, which was still the frontier in the late 1800s, many log buildings appeared as new settlers found land in the cutover.

Notching the corners of log buildings required considerable skill, and ethnic log builders had their own special approaches. This Norwegian log barn illustrates the carefully formed corners. Waukesha County, Old World Wisconsin, Eagle.

Many of the early log barns were left unchinked except during the coldest months of the winter because the farmers believed livestock needed fresh air to prevent sickness, a belief that was not altogether inaccurate. During the coldest weeks in the winter, the cracks between the logs were stuffed with hay and straw to keep out the frigid winds and prevent the snow from sifting through the walls.

The Germans built their log barns with spaces between the logs. These spaces were immediately chinked and then recaulked as the wood dried and shrank. They used clay, rye straw, and lime plaster as chinking material.

The Norwegians built their log buildings with snug, tight-fitting joints requiring a minimum of chinking. They cut the logs square with their axes so they fit closely together.

The Finns did something different. They built two basic types of log barns: one to house their livestock, another to store their hay. Rather than store hay near where the livestock were housed, the Finns often built hay barns in their hay fields. They reasoned that if the main buildings burned, they'd still have a supply of hay. These hay barns were built of round logs, and no attempt was made to fit them tightly together (see page 149). The spaces between the logs allowed for a free exchange of air so the hay would cure. For their livestock barns, the Finns squared the sides of the logs, left the tops round, and hollowed out the bottoms so they fit snugly together.

Several other ethnic groups also built quite distinctive wood-frame barns. To some extent it is possible to identify these ethnic barns in Wisconsin, but in most cases it's difficult to detect ethnic influences because the building choices that proved most practical and efficient were widely adopted and continued, while less-useful approaches were forgotten, whatever their origins.

## ENGLISH BARNS

One of the simplest wood-frame barns in Wisconsin was the English barn. Earlier, this little barn had been introduced to New England by various nationalities, but primarily the English. The barn was usually 42 feet to 60 feet long and 28 feet to 30 feet wide. The barn did not have a basement and often was made of wood from top to ground level.

Some refer to the English barn as a Connecticut barn or a three-bay barn. The latter term comes from the barn's basic structure—a haymow on either side of a threshing floor. In an architectural sense, the word *bay* goes back more than two thousand years. It relates not to a bay window but to spaces between structural members, such as the spacing of arches in the cathedrals of the Middle Ages.

Some of the earliest barns in Wisconsin were three-bay threshing barns, used to store wheat prior to threshing. These were the days before grain binders and threshing machines, when the grain was cut with a cradle and threshed with a flail on the threshing floor of the barn. Waushara County, Highway 73 north of Neshkoro.

The English barn had doors on the long side leading to the threshing floor, and it had a gable roof. The majority of these barns were constructed with the siding boards placed vertically. In addition to providing a place to thresh grain and store hay, many English barns were also built to house livestock. At the ground level of one bay were the livestock, and above them was a floor piled with hay (which also served as insulation for the animals). On the other side of the threshing floor, hay was piled from ground level to the roof rafters.

## Scandinavian Barns

The Norwegian frame barns in Wisconsin, following the early log ones, had strong European influences. The gable roofs were long and steep. It is likely the early ones had been thatched with straw or sod; the later ones were roofed with cedar shingles. An extension at the gable ends of the barn supported the hayfork track. The long, steep roof that came within a few feet

Norwegian log barn at Old World Wisconsin, Eagle.

**Finns built some of the finest log buildings in the state. They meticulously hewed and fit the logs together so that little or no chinking was necessary. Waukesha County, Finnish barn at Old World Wisconsin, Eagle.**

of the ground and the extension jutting out at the gable ends gave the barn a distinctive and most attractive appearance. This barn's shape is somewhat similar to modern A-frame buildings, but its roof is not quite as steep as the A-frame's.

At first impression, a Finnish barn looks too tall for its length and width. These barns were often no larger than 28 feet by 40 feet. The Finnish barn was a two-story construction, with cattle housed on the ground floor and hay stored above. A hay door opened at the gable end, where hay was pulled by a hayfork into the haymow.

The Finns sometimes built wood-frame barns that combined log and frame construction. An interesting example is a Finnish barn hauled from northern Wisconsin to Old World Wisconsin, a museum of historic ethnic architecture near Eagle. This is a two-story barn, rectangular in shape, with a place for livestock on the ground floor and hay storage in the upper part. The stable part of the barn is of log construction; the hay-storage area is wood-frame construction. The barn's steep gable roof is shingled with cedar shingles and topped with a small, simple cupola. The logs in the barn are squared and dovetailed together at the corners. One interesting element is the height of the barn doors: they are less than six feet high, which suggests the Finnish farmers were not tall.

# German Barns

Many of the wood-frame barns built by Germans in the late 1800s and early 1900s had nothing peculiarly German about them. Many of the bank barns described later in this book (see page 33) were built by German farmers. Some Germans built barns following a technique called half-timbering, however, and these barns are distinctly German. Relatively few half-timber barns were built in the United States, and those known to remain are found only in Wisconsin, Pennsylvania, Missouri, and Texas.

Half-timber barns were constructed of heavy timbers, usually white oak, that were mortised, tenoned, and then pegged together. No nails were used in building the barn's frame. The spaces between the timbers were filled with nogging to form the walls of the barn. Nogging could be one of several things—kiln-fired brick, air-dried brick, rubble masonry, clay and straw on wood slats, even stovewood. Stovewood (see sidebar below) was probably the most unusual nogging used. The roof boards of a typical half-timber barn were from twelve inches to twenty inches wide and were covered with hand-split shakes. Some of the early half-timber barn roofs probably were thatched, as was customary in some areas of Europe.

---

## ⚜ Stovewood Barns ⚜

**Stovewood, ordinary cordwood of the type used in wood-burning stoves and fireplaces,** was used as nogging in some half-timber construction. Stovewood was also used to build entire barn walls. Stovewood buildings are a relative rarity. One can find them primarily in Door County and some other northern and western Wisconsin counties. It is not known if they originated in Europe—some believe the rare stovewood barns in Norway may have had their roots in the United States. As to ethnic connection, apparently several ethnic groups, including Polish, Norwegian, and Finnish, built them. Though used in some half-timber construction, stovewood is not usually associated with half-timbering, and it is clearly not always of German origin.

When stovewood was used as nogging in half-timber construction, the thickness of the timbers determined the length of the stovewood pieces in the wall. Timbers were usually seven inches to eight inches thick, and the stovewood was cut so it was flush on both sides of the timbers. Sometimes the stovewood was coated with lime plaster, but often boards were nailed over the walls.

One of the first stovewood buildings discovered in Wisconsin was a house near Williams Bay, in Walworth County. Some clapboards fell from the house, revealing the stovewood walls underneath. Research indicated the house had been built in 1848 by David Williams from New York. This house was built entirely of stovewood, without incorporating the half-timbering type of construction. The wood for this house, entirely oak, had been cut, sawed, and split into sticks fourteen inches long. It probably came from the nearby woods. About twenty thousand sticks of wood had been piled one on top of the other and mortared together.

Today, many of the remaining stovewood buildings, including the stovewood barns, are located in the northeastern part of Wisconsin. Rare examples are also found in Oneida, Bayfield, and Adams Counties.

Interestingly enough, the stovewood barns of the Door Peninsula are of the half-timber type, whereas those west of Green Bay—not the city but the bay—have solid stovewood walls. On the Door County side, the stovewood used was primarily cedar or pine; across the bay, hardwoods were used about as often as softwoods. Sometimes softwoods and hardwoods were used together, probably reflecting what was available to the builder. The stovewood buildings on the Door Peninsula date from roughly 1890 into the early 1900s. Those across the bay were built after 1900. This type of construction on both sides of the bay appears to have peaked about 1910.

Barns constructed of stovewood are relatively rare in the United States. A cluster of them was built in Wisconsin, Upper Michigan, and several other north-central states. Various ethnic groups built them in places where rot-resistant cedar and tamarack wood were prevalent. This stovewood Polish house barn housed animals on the left and the farm family on the right. Waukesha County, Old World Wisconsin, Eagle.

In constructing a solid stovewood barn, lengths between eight inches and twelve inches were used. Sometimes the builder split the wood randomly; sometimes it was squared; sometimes the wood was left round. The stovewood was usually imbedded in lime or lime-and-cement mortar in a fashion not different from how fieldstone walls were constructed. For the corners, either short, squared logs were laid parallel to the wall's plane or a squared post was stood vertically to the height of the wall.

In barn construction, stovewood was often used to add a wing to a barn constructed earlier of logs. One of the reasons the stovewood approach to construction had some popularity in Wisconsin was the ease with which such a building could be put up. One man and his family could construct a stovewood addition to their barn without bringing in outside help. Such an addition could be put up in a few days, compared with the several weeks required to erect another type of barn.

Scattered examples of stovewood buildings have been found throughout the north-central states, especially Upper Michigan and Wisconsin. Because the stovewood is often covered with another material, a stovewood barn often is not found until it is being torn down or is in such disrepair that the siding begins falling off the walls. Then it is usually too late to save the building.

The study of pioneer barns thus requires some careful detective work. The pioneer's log house often became a part of the barn when a frame house was built, or it was hidden within the new house. Both log and stovewood buildings were often covered with siding, giving them, at first glance, the appearance of buildings constructed of more modern materials and in a more modern way. The student of pioneer barns must literally look beneath the surface of a building to determine its history, especially in the case of log and stovewood construction.

This early German half-timber barn has a thatched roof and overhang. Waukesha County, Old World Wisconsin, Eagle.

In a half-timber barn, the timbers were visible in the outside walls. The spaces between timbers were filled in with the various materials already described. This was in contrast to a wood-frame barn, which had outside walls of wooden boards and no timbering visible to the outside.

The early German immigrants brought with them craftspeople skilled in carpentry, joinery, cabinetmaking, stonemasonry, and blacksmithing. Many of them were familiar with the half-timber form of construction, known in German as *Fachwerkbau*. From about the Middle Ages on, half-timbering had been quite a popular building approach in much of Europe. In addition to barns, these early German settlers in Wisconsin constructed half-timber houses and churches.

As various ethnic groups arrived in Wisconsin, they brought with them knowledge of buildings they had known in their home countries. Waukesha County, German barn at Old World Wisconsin, Eagle.

The bank barn is one of the most popular barn styles in Wisconsin. Typically the barn was built against a hill (or bank) to allow easy access to the hayloft for unloading hay. The stable was located in the lower part of the barn. Monroe County, Highway 33, Town of Portland.

# 3

# POPULAR BARN STYLES

## *Bank Barns and Variations*

Of the various styles of barns in Wisconsin, bank barns are the most prominent. They are found in all parts of the state, wherever dairy cattle were raised. The bank barn is a two-story structure as high as 60 feet or more, often 34 feet or more wide, and sometimes more than 100 feet long. The first floor of the bank barn is where the cattle are housed; the second floor is for the storage of hay and sometimes grain. The walls of the first floor, or stable, of the bank barn are traditionally constructed of fieldstone, quarried rock, poured concrete, or concrete block. Fieldstone and quarried rock are found in the older barns. Most barns built after about 1920 have walls of poured concrete or concrete block.

The wooden frame of the bank barn sits on the stable walls. Usually the stable is partially underground. By having the barn built against the side of a hill or bank (thus the name for the barn), a farmer could drive directly to the second-story threshing floor with a load of hay. A bank barn is an English barn, placed on a raised foundation so livestock can be stabled underneath the hay storage space. Depending on the length of the barn, there can be two or even three threshing floors. Each one has its own set of double doors reaching from floor to eaves— often 12 feet to 16 feet high. The threshing floor is the heart of the second story of the bank barn. In the early days, before threshing machines, grain was threshed there. In later years, it was used mainly for the unloading of hay into the cavernous haymows on either side.

Before the corn shredder and corn picker became widely used for harvesting corn, many farmers used the threshing floor as a place to husk corn. After the corn had been cut and shocked and had had time to dry, it was hauled to the barn. Farmers spent long hours working on the threshing floor, stripping the yellowish-white husks from the golden-yellow corn by hand. The threshing floor was not heated in any way, except for the heat that rose from the stable below, but it protected the farmer from wind and snow, if not from cold.

The threshing floor served other purposes, too. Many a farmer drove a load of fresh hay there just before the first drops of a summer thunderstorm pelleted down on the barn roof. Tons of hay were thus spared a soaking. During the winter months, the threshing floor, if it wasn't piled high with hay that wouldn't fit into the mows, became a place to store stray pieces of machinery.

The threshing floor also provided a place for farm boys and girls to practice their acrobatic skills. Every barn had a system of ropes and pulleys used to lift the hay from the hay wagon into the mows. What great fun it was to swing from one haymow to the other, tightly grasping the hayfork rope. Of course this activity could be more than a little dangerous if the threshing floor was bare. Usually, though, at least a small pile of hay or straw was there to serve as a safety net for the unfortunate youngster who misjudged and found him- or herself swinging high, not quite able to reach the haymow on the opposite side and without enough momentum to return to the mow. Dropping twenty feet into the pile of hay was often as much fun as successfully swinging the distance from mow to mow.

A bank barn's stable usually faced south or east and was at ground level, allowing easy access for the cattle. Lunde farm, Dane County, Town of Christiana, circa 1874. WHi Image ID 26142

Other athletic feats youngsters performed on threshing floors included grasping a spiny, hemp hayfork rope and trying to scale it to the hayfork track bolted just under the peak of the roof. Any farm kid who couldn't "climb the rope" was seen as some sort of weakling.

This bank barn's large doors open to the hayloft. The farmer could drive a team of horses and a load of hay into the barn through these doors and unload the hay into haymows on either side. Fond du Lac County, Oaklane Road and County Road F.

A favorite trick played on city cousins was to encourage them to climb the rope without any instruction about how to return to the floor again. The city cousin would climb the rope and then, when his or her arms got tired, slide back to the floor, suffering rope burns to the hands because the country cousin had conveniently forgotten to demonstrate how to wrap the legs around the rope to slow the descent.

Several variations of the bank barn are found in Wisconsin. The traditional Pennsylvania bank barn, found in southern and eastern Wisconsin, and particularly in Green and Sheboygan Counties, can be traced to Europe. It is found in the highlands of Upper Bavaria, the southern spurs of the Black Forest Mountains, the Jura region, and elsewhere in Switzerland.[1]

During the seventeenth, eighteenth, and nineteenth centuries, Swiss and Germans and several other nationality groups settled in Pennsylvania and brought with them their knowledge of barn building. The Pennsylvania barn thus became a merger of Swiss and German influences. The immigrants to Pennsylvania changed the European bank barn in one

The first Swiss immigrants began arriving in Wisconsin in the 1820s, before Wisconsin had become a territory. By 1900 about twelve thousand Swiss lived in the state, most of them in southern and southwestern Wisconsin, and especially in what is now Green County. Many of these Swiss immigrants were farmers who built barns such as this one, with elaborate weather vanes and cupolas, and occasionally special rounded windows. These farmers and the Swiss cheese makers helped make Green County the onetime Swiss cheese center of the country. Freitag barn, Green County, about one and a half miles north of Monticello. **WHi Image ID 40643**

The pentroof, an overhang above the stable windows, provided a place for cattle to get out of the weather. Pentroofs are often associated with Swiss barns. Green County, Highway 62 and Old Highway 62 north of New Glarus.

important way. Though both the American version and the European version relate to the terrain, the European barns were built at right angles to the lie of the hills. In this country, the barn was set parallel to the hill.

In addition to the basic features of a bank barn—a threshing floor, two stories, built against a hill—the Pennsylvania bank barn features a forebay. The forebay is an extension of the upper part of the barn. It extends beyond the stable wall on the side of the stable that is exposed. In bank barn construction, one side of the stable is protected by the side of the hill against which the barn has been placed. The forebay thus provides some protection for the doors and windows on the exposed side and allows a place for the cattle to gather out of the wind and weather. The bank barn usually faced either south or east, and, with the forebay extension, cattle could be quite comfortable outside even on the most blustery days.

A modification of the forebay is the pentroof. Some barn builders, rather than building the forebay extension, constructed a little roof over the doors and windows on the exposed side of the barn. This also afforded protection for the cattle on stormy days. Eric Sloane, a New England artist and writer, explains that the pentroof really should be called a penthouse. *Penthouse* comes from the old French word *apentis,* meaning "a small roof attached to a building." Common usage has changed the word *penthouse* to mean something quite different. In barn language, *penthouse* has been changed to *pentroof,* not altogether correct but commonly accepted.[2]

Many barns in Green County closely resemble the traditional Pennsylvania barn. When we think of Green County, we usually think of the Swiss immigrants from Canton Glarus who settled in the area. But according to architectural historian Richard W. E. Perrin, the first settlers in the county were Pennsylvania Germans, who brought with them their knowledge of Pennsylvania barn building.[3] Another characteristic of the Green County Pennsylvania barns is the location of several round-topped, vertical-louvered ventilators in the wood-frame walls of the upper part of the barn. At first glance, these ventilation louvers look like windows in the side walls of the barn.

Many versions of the bank barn exist around the state. The most common variation from the traditional Pennsylvania barn is the elimination of the forebay and the pentroof. Another variation is in roof type, the subject of chapter 5.

The earlier bank barns, those dating back to the 1880s or before, usually had hewn beams—beams formed with a broadax. Most barns after the turn of the century included sawed beams. The barn beams were often as large as 12 inches by 12 inches and were pegged together with wooden pegs rather than nailed together. The superstructure was usually covered with vertical boards, often a foot or more wide and 16 feet long. Because a portion of the stable was underground, some farmers walled off part of the stable and used it as a root cellar. This served well, as the portion of the stable totally underground was protected from frost. During extremely cold weather, a door that connected the stable with the root cellar could be left open to add heat to the root cellar.

**When a natural hill or bank was not available, farmers constructed earthen ramps to allow access to the hayloft. These barns were also considered bank barns. Jefferson County, Highway 106, approximately three miles east of Highway 73.**

Popular as it was around the state, the bank barn was not without its critics. Some farmers claimed, and with considerable evidence, that the bank barn was dark, damp, and dingy. They felt it was not healthy for animals to be housed in such circumstances or for farmers to work under such conditions. J. H. Sanders, in an 1893 book on barn building, wrote:

> The dark, cavernous recesses of very large barns are seldom ventilated or dry, and this is almost necessarily a consequence of great size. The heavy foundations imply a basement dark, damp, and stinking. The great roof and floors mean heavy timber, much skilled labor, and expense, and last but not least is the chance that some winter night, in red and yellow flame skyward soaring, the huge structure vanishes with all the horrors of agonizing death to helpless creatures . . . and to the owner loss immeasurable. Let us rather have two or more smaller buildings, all above ground, on light foundations, light timbers, but little "framing"—the "balloon" style of construction permitting the use of much ordinary labor—far enough apart for some degree of safety from fire and a chance to save life. Doors on every side and ample windows. Nothing is so cheap as sunlight and yet nothing is so scarce in the average barn or stable.[4]

Some rather obvious questions anyone interested in Wisconsin barns might ask concern these massive barns. Why do we find so many bank barns and other huge barns in Wisconsin? Why were they built? And why were most of them built during the late 1800s and the early 1900s?

Before the Civil War, few of the big barns could be found in Wisconsin. Granted, most of the northern parts of the state had not yet been settled by the Civil War era. But even in the southern and central parts of the state, which were quite well settled, few of the big barns could be found. To answer the questions, we must look back to the beginnings of agriculture in the state and see how it changed dramatically following the Civil War.

The first farmers to arrive in Wisconsin came from New England and the eastern states, where they had previous experience growing wheat. When the immigrants began arriving from Europe, they quickly followed the lead of the Yankees and began to grow wheat. By the Civil War, Wisconsin was a leader in the nation in the production of wheat. When settlers arrived in Wisconsin they immediately broke the soil and planted wheat, and each year they tried to break additional acres so they could grow more. These early farmers were commercial farmers, not subsistence farmers. They farmed to make money, and wheat was the way to do it.

Wheat farming did not require large barns. In fact, many farmers had a barn only large enough to house their work animals. There was no need for the great barns we find today. But there were problems with wheat growing. Farmers had to cut the crop by hand, using the ancient method of cutting it with a cradle—a scythe with a wooden tooth attachment. One man with a cradle could cut only two to three acres of the golden wheat in a day. It took so long to cut the crop that by the time a farmer got to the last of his acres, much of the wheat was overripe and had shattered to the ground and been lost. And threshing, too, was an ancient process, done by either beating the grain with a flail or walking over it with oxen.

As wheat growing flourished in the state, several people worked on the problems wheat farmers faced. In 1834 Cyrus McCormick patented a reaper, a machine that would cut the wheat and do the work of several men. Other reaper inventors followed, including George Easterly of Walworth County, Wisconsin.

In 1844 J. I. Case of Racine brought to the market a mechanical threshing machine that separated wheat kernels from the stems. Now the threshing operation could be speeded up many fold, and the job was far easier. With both the cutting problem and the threshing problem solved, nothing could hold back the expansion of wheat growing in Wisconsin. The number of wheat acres increased. In 1859 Case came out with a more modern threshing machine powered by a ten-horse sweep. This machine became the standard for threshing wheat in the state. Still more acres of wheat could be grown.

From early settlement days until after the Civil War, Wisconsin was a major wheat-growing state. Before the mechanical reaper arrived, farmers harvested their wheat fields with cradles. **WHi Image ID 26284**

Wisconsin produced nine million bushels of wheat in 1855, twelve million bushels in 1856, and fourteen million in 1857. It is estimated that in 1860 between twenty-seven million and thirty million bushels of wheat were grown in Wisconsin. And it sold for eighty cents a bushel, compared with the forty-five or fifty cents received several years earlier. Wheat was king. Farmers paid off their debts and argued with anyone who dared suggest that wheat farming wouldn't forever be the principal source of farm income in Wisconsin. Everywhere Wisconsin was mentioned, wheat was mentioned soon after. Who could argue with success? However, by 1860 a few people began to wonder about wheat and the wisdom of such total dependence on it.

With the passing of the wheat era in Wisconsin, farmers turned to other cash crops, such as tobacco. The building at right is a tobacco barn, or tobacco shed, used for drying tobacco leaves. For many years Wisconsin was a leader in producing cigar wrapper tobacco. Vernon County, West Old Town Road west of Westby.

Then it began to happen: poor weather descended—too dry to ensure a good crop, too wet when it was time to harvest. Diseases appeared in crops that were repeated on the same fields year after year. Smut and rust began to take their toll. Adding to all of these problems was the lowly little chinch bug, a grayish black insect with a light X on its back. With its mosquito-like beak, it drilled into the stalks of the wheat and sucked out the life-giving juices. The wheat crop wilted and died. Millions of chinch bugs moved into Wisconsin wheat fields. All the farmers could do was curse them as they watched their sole source of income dry up and rattle in the hot, dry winds of August that blew across the land.

Farmers had no way of knowing it, but the bumper crop of 1860, the biggest ever, was never to be surpassed. In fact, wheat yields never came close to that record again. By 1879 wheat growing in Wisconsin was essentially dead. Farmers were in turmoil. What could replace this steady crop that had served them so well from the days when they first turned the soil with a breaking plow?

Wisconsin farmers searched frantically for a way to make a living. Some packed their belongings and moved farther west, where wheat growing could be tried again. Many of the settlers who had come from New York, Pennsylvania, and New England chose this option and moved on to Minnesota, the Dakotas, and other western states, hoping to escape the chinch bugs, the diseases, and all the other problems that had turned wheat farming in Wisconsin into a nightmare.

But many farmers stayed in Wisconsin. They were committed to a life of farming in this state and hoped they could, prayed they would, find a way to feed their families on the land. Tobacco growing was introduced into the state as early as 1840 and grew to some prominence in southern and western Wisconsin as a cash crop, where it continues to this day in a much-limited way. Several other wheat substitutes followed a boom-bust pattern. During the Civil

War, sorghum was in demand as a sugar substitute, and sorghum growing briefly flourished, but after the war its importance decreased. The growing of hops was introduced into the state about the same time. For a while, hops became a very popular crop and commanded high prices. But the bottom dropped out of the hops market in 1869, and production fell off in response. Sheep raising became popular in many of the southern counties, gaining prominence during the Civil War, when raw wool earned a good price. But after 1870 wool prices fell, and sheep producers became discouraged.

In central Wisconsin, on the sandy soils of Waushara, Portage, Waupaca, Marquette, and Adams Counties, the farmers shifted from wheat to potatoes. Wheat had been the primary crop in the area ever since it had been settled, much of it after 1850. By the 1880s the same problems that had struck southern Wisconsin wheat farmers visited central Wisconsin. Potato growing seemed the answer. Essentially no new farm machinery was needed. Potatoes were planted with an inexpensive hand planter, cultivated with a horse-drawn cultivator, and harvested with an ordinary six-tine barn fork.

**In central Wisconsin, potatoes became a valuable cash crop in the late 1800s and early 1900s. While waiting for higher prices, farmers often stored their crop in potato cellars like this one until late winter. Potatoes were stored in the lower part of the building; farm machinery was stored above. Waushara County, County Highway A and 15th Road.**

In 1894 Waushara County grew eighteen thousand acres of potatoes, yielding nearly two million bushels. In 1910 more than two million bushels of potatoes still were being grown in the county. Storage, however, was a problem. Wheat could be stored without any fear of its freezing, but not so with potatoes. They were a perishable crop and had to be protected from freezing temperatures. Potato warehouses sprang up in many villages on rail lines. Hancock, a Waushara County village of about six hundred, had seven warehouses in 1915. By that same year, Wild Rose, also in Waushara County, boasted eight potato warehouses.

A man tends to the daily milking near Green Bay, circa 1895. WHi Image ID 60973

King potato had dethroned king wheat in the central-Wisconsin counties. But during the 1930s, with decreased yields, poor weather, and low prices, potatoes suffered the same fate as wheat. (Today, with modern irrigation practices, thousands of acres of potatoes continue to be grown in central Wisconsin on the sandy soils where historic yields had been unpredictable because of the weather.)

When wheat farming began to die, some farmers became more serious about hog production. Whereas during the pioneer days the hogs ran loose in the woods, foraging for acorns and whatever else they could find to stay alive, now the farmers planted corn and fed it to their hogs. Pork sales could help to pay off the mortgage—or at least keep them from losing their farms.

But it was the dairy cow, the lowly milk cow that the Wisconsin farmer had ignored for years, that slowly became the economic mainstay in Wisconsin agriculture. The transition from grain farming to dairy farming did not happen quickly, nor without considerable controversy.

Dairy farming demanded a different lifestyle for the farmer who had grown up with wheat. Did the farmer, the tiller of the soil, the free spirit on the land, want to become tied to a milk cow? The dairy cow demanded much from the farmer. For 365 days a year, both morning and night, the cows had to be milked. They had to be fed daily, and the manure had to be removed and hauled back on the land. Did the Wisconsin farmer want his entire life to revolve around the lowly milk cow? For years, it had been woman's work to care for the cows. What would it do to his masculinity if he, a man of the soil, started doing woman's work?

Dairy farming presented another problem. How would the dairy products be marketed? For many years, it had been common for the farmer's wife to skim the cream from the excess milk produced by the two or three cows the farmer owned, churn the butter, and then trade it with the storekeeper in town for groceries. The quality of this butter ranged from quite good to not fit for human consumption. Some of it could be used only as wagon grease. Wisconsin butter had a poor reputation in the cities where it was marketed, such as New York, Philadelphia, St. Louis, and Chicago.

Chester Hazen of Ladoga, in Fond du Lac County, is credited with opening the first commercial cheese factory in the state in 1864. Making cheese required skills that few farmers possessed. Yet cheese was a product that was less perishable than butter, and one for which there was a market demand. By 1870 ninety cheese factories could be found in Wisconsin, with the number increasing each year. But the advent of the cheese factory in Wisconsin presented a philosophical problem to the Wisconsin farmer. The cheese factory was the first step toward the industrialization of agriculture. The farmer could see himself becoming one cog in the system, that of producer of raw materials.

The farmer faced a challenge to his basic values. He had to adjust not only to doing woman's work and to the demands of the dairy cow but also to the demands of the cheese factory, which insisted on certain levels of quality and told him when to deliver his milk. How much control was the independent farmer willing to accept? Did he want to become part of a new industrialized agriculture? As it turned out, he had little choice. He needed a market for his dairy products, and the cheese factory provided it.

The cream separator, invented by Carl Gustaf De Laval in 1885, influenced dairying in another way. Many farmers purchased cream separators because they could easily separate the cream from their milk and then market the cream at creameries, which churned butter and were becoming popular in the state.

The Yankees who settled in Wisconsin, particularly the New Yorkers, deserve credit for providing leadership in starting the dairy industry in Wisconsin. Though they had been wheat farmers in Wisconsin, many of them had experience with dairy farming in their home state. It was often New Yorkers who headed local movements to build cheese factories and organize breeders' associations to promote the dairy industry in Wisconsin.

W. D. Hoard came to Wisconsin in 1857 from New York State, where dairying was popular, and became one of these dairy leaders. Through his newspaper, the *Jefferson County Union*, he wrote about dairy farming and its virtues. He advocated better storage facilities for butter and cheese made in the summer. He suggested ways of reducing the cost of winter feeding of cattle. He advocated the building of silos. And, probably most important for the future of dairying, he preached that farmers should breed their cattle for milk production only and give up on dual-purpose cattle. Dual-purpose cattle, bred for both milk and meat, seldom did very

As dairy farmers grew their operations, they added on to their barns. Many farmers also continued raising other animals such as hogs and beef cattle. The addition on the right of this barn was likely for hogs. Dodge County, County P south of State Road 60.

well in either area. In 1885 Hoard founded *Hoard's Dairyman,* a magazine that continues to this day to provide up-to-date information about dairy farming.

The University of Wisconsin's College of Agriculture, a fledgling school groping its way in the late 1800s, became interested in dairy agriculture and dairy farmers. In just a few years, the college became a national leader in dairy research and the extension of new information to the dairy farmer. Several men contributed to the success of the new agriculture college. Hiram Smith, a Sheboygan dairyman, was appointed to the board of regents in 1878. He promoted the appointment of William A. Henry to a professorship in agriculture in 1880. Henry became dean of the College of Agriculture, which was created in 1889, and even though trained as a botanist, he soon became a solid promoter of the dairy industry in Wisconsin.

The University of Wisconsin had many prominent early researchers, including Stephen Moulton Babcock. Dean Henry asked Babcock to develop a milk test, and soon the "Babcock Test" provided a simple, reliable way to determine the butterfat content of milk. Thus, a mechanism for equitably paying farmers for the milk they produced was established. Professor Franklin H. King researched the silo as a way to store corn for winter feeding. Professor Harry Russell introduced a test to determine if cattle had tuberculosis. And Professor Henry's handbook for stockmen, *Feeds and Feeding,* became a classic in the area of dairy cattle nutrition.

Wisconsin had become a dairy state by 1900, but the transition from wheat farming had taken nearly fifty years. Becoming a dairy farmer, unlike trying a new cash crop for a year or two, was a nearly permanent decision. Cattle had to be purchased; buildings had to be built. When a farmer went into dairying, he committed himself to a type of agriculture he couldn't easily leave . . . which leads directly to the big barns. To become a dairyman, the farmer had to house his cattle in a structure that would protect them from the long, cold Wisconsin winters. If he wanted a steady flow of milk during the winter months, the cattle couldn't be allowed to ramble around the barnyard seeking shelter under a straw shed. They had to have a weatherproof structure.

For two thousand or three thousand dollars, a farmer could build a barn to house cattle, but that was a considerable investment. When a man decided to go into dairying, he wanted a building that would be substantial and permanent. Though it may have taken a farmer ten years to decide to go into dairying, when he made up his mind, he decided not only for himself but for his children as well.

It would be dramatic to say that the lowly chinch bug that sucked the juices from the wheat stems built the big barns in Wisconsin. But such a conclusion is simplistic. As anyone familiar with crop growing knows, repeated planting of the same crop in the same fields often leads to problems with diseases and pests. It was inevitable that wheat farming could not continue year after year in Wisconsin but needed to be replaced with a more diversified form of agriculture.

Dairy farming became that more diversified agriculture, and the big barn became the focal point of dairy farming in the state. It remains as the visible symbol of a great transition in Wisconsin agriculture. And many of the barns built in those early days of Wisconsin dairying are still in use today.

## OTHER TWO-STORY BARNS

Scattered throughout the state are large two-story barns, some of them built on top of masonry basements, some not, and many equally as large as the bank barns described earlier. These barns do not have threshing floors, and thus one cannot drive a wagon inside to unload hay. Instead, hay is unloaded at large doors at one or both gable ends outside the barn. Usually the hayfork track extends a few feet out from the barn, supported by metal brackets or a wooden extension on the end of the barn roof.

Technically the Norwegian and some of the Finnish barns described earlier (see page 24) fall into this category. But a large number of two-story barns have characteristics different

As technology developed for moving hay inside the barn, the possibility of two-story non-bank barns emerged. The hayfork track was extended from the peak of the barn so hay could be unloaded outside of the barn and carried inside through a system of ropes and pulleys to the barn's hayloft. Monroe County, Highway 27, near Cashton.

from those of the Norwegian and Finnish barns, and these barns are discussed later. The primary difference is size. These two-story barns are larger than the ethnic barns described earlier.

Compared with bank barns, these barns were less expensive to build, for it was not necessary to construct a threshing floor strong enough to support a team and a wagon full of hay. Without a threshing floor, the barn had a clear space the entire length of it to store hay. Also, because no bank was required for access to the threshing floor, the barn could be located anywhere on the farmstead. The first floors of these barns were constructed of various materials—some of fieldstone, some of poured concrete, a few of masonry blocks (particularly those constructed more recently). Some of these barns were constructed of wood from top to bottom, including the first story.

This type of barn does not appear to be clustered in any one part of the state. In the same neighborhood may exist two-story barns of this type and bank barns. Such factors as cost, the personal interests of the farmer, the skills and inclinations of the barn builder, and the terrain probably influenced which type of two-story barn was built. The building of two-story non-bank barns in the 1940s and 1950s probably resulted from the development of mechanized hay-handling equipment. Bales of hay could be elevated from a wagon outside a barn to a conveyor system inside the barn. Thus hay could be moved easily to all parts of the barn's storage area.

---

### ✥ Village Barns ✥

**From the time of the first settlements in Wisconsin until the coming of the automobile,** behind many village homes was a barn. These obviously were not the big barns being built on the farms throughout the state, but they did serve a most useful purpose.

Until the automobile became popular, village people—country folks, too, of course—depended on the horse for transportation. Many villagers also kept a cow or two to provide milk and butter. Sometimes a lean-to chicken house was added to the village barn; occasionally the barn was divided to provide space for a small flock.

Village barns often were not much larger than an unattached garage, but they were almost always two stories. The majority of the village barns standing today have been converted into garages. Driving through any small town in Wisconsin, one can easily pick them out. They were usually simple rectangular structures with gable roofs and double doors on the front to admit carriages.

Some of the early barn plan books provided suggestions for building village barns. For example, William A. Radford included in his plan book a "Small Barn for a Village Lot—Cost of Blue Prints, $3.50." The barn described is 18 feet by 20 feet and is designed to accommodate four horses and a wagon or carriage. The loft on the second story stored three or four tons of hay.

The construction of this little barn is about as plain and simple as it could be and still have it look right when finished. Nobody likes a cheap looking building, but no one objects to a good-looking building if they get it cheap. The problem is how to build what will be satisfactory in a few years' time. Sometimes an inexpensive building may be shaded with trees or screened by vines in such a way as to give it a presentable appearance even in winter. An evergreen or two planted along the side, if there is plenty of room, makes a great winter addition to the looks of a stable. Grape vines usually do well if suspended by wires from the eaves, but grape vines should never be tacked close to the side of a building, they need air on all sides.[1]

It's interesting that Radford was so concerned about "good looks" for his barn, in addition to low cost.

Some very elaborate village barns were constructed. These had storage places for wagons and carriages and housed several horses and a cow or two. The most elaborate also provided living quarters above the stable for the servants and other hired help employed by the well-to-do village dweller. But this type of barn was clearly the exception in Wisconsin.

On the farm, the horses were usually housed in the same barn with the cattle. Occasionally a separate horse barn was built. The carriages were usually stored in a separate building along with other farm machinery. The village barn put all the means for transport under one roof—horses, carriages, and the necessary feed. The village barn is slowly passing from the scene as new homes are built or people feel the need for more up-to-date garages.

During the years when horses still provided transportation for many people, it was not uncommon to find a small barn next to a village home. Typically, a horse or two were housed in the lower portion and hay was stored above. Waushara County, Park Avenue, Wild Rose.

**Notes**

1. William A. Radford, ed., *Radford's Combined House and Barn Plan Book* (Chicago: Radford Architectural Co., 1908), 211.

The twenty-sided Nashold barn is one of Wisconsin's most unusual barns. Built in 1911, it has been restored and is listed on the National Register of Historic Places. Columbia County, County Highway Z, .4 mile east of State Highway 146, Town of Fountain Prairie..

# 4

# AN ALTERNATIVE FORM

## *Round and Polygonal Barns*

Round and polygonal buildings trace back to ancient Greece and Rome, years before the birth of Christ. Through the years some interest in centric buildings continued in Europe, and the immigrants brought these ideas with them to the New World. In this country, round and polygonal buildings attracted attention in the 1850s, largely due to the work of Orson Fowler, a phrenologist from New York.

## THE OCTAGONAL FORM

Fowler became interested in architecture, particularly in the octagonal form. In his 1854 book, *A Home for All,* Fowler praised the virtues of octagonal buildings. He wrote, "Nature's forms are mostly spherical. She makes ten thousand curvilinear to one square figure. Then why not apply her forms to houses? Fruits, eggs, tubers, nuts, grains, seeds, trees, etc., are made spherical in order to enclose the most material in the least compass."[1]

Fowler's penchant for the octagonal form extended well beyond the home. He made specific suggestions about building octagonal barns. In an octagonal barn, he said, cows should face the center in order to save the farmer steps. The octagonal barn should be built into a bank or knoll so the main floor could be reached with a wagon and the lower floor should be relegated to animals.[2] This, of course, makes the octagonal barn a type of bank barn.

Build the barn big, Fowler said, so there will be plenty of room for livestock and feed storage. The farmer should construct as few buildings as possible, putting as much as could be under the barn roof. Ideally, Fowler recommended, the farmstead should consist of one barn and no buildings around it, one house and no additions to it. Of course, both the house and barn should be octagonal. He suggested incorporating glass in the roof of the barn so the building would have plenty of light.[3] This suggestion, quite radical for its time, reflects Fowler's insight into a problem characteristic of the larger octagonal barns as well as round barns—that of providing enough light in the center parts of the structure.

Octagonal barns are among the more unusual barn styles found in Wisconsin. Few remain standing. Sauk County, Highway 23 near Plain.

Ernest Clausing, a German carpenter, built octagonal barns, many of them in Washington County in the 1890s. Old World Wisconsin's restaurant is a Clausing barn that was moved to the site in Eagle.

Though many of Fowler's suggestions seem to have been followed in the construction of Wisconsin's octagonal barns, it's difficult to believe his ideas for barn building were followed directly. Fowler visited Wisconsin in 1850, proclaiming the virtues of octagonal-shaped buildings, and he obviously influenced many Wisconsin people to build octagonal houses, evidenced by the Milton House in Milton (known for its connection to the Underground Railroad), the headquarters for the Watertown Historical Society, and the historic octagon house in Fond du Lac. But it wasn't until the late 1800s and early 1900s that octagonal barns were built in Wisconsin, more than forty years after Fowler's other ideas were at their peak in the country. The largest single grouping of octagonal barns in Wisconsin was in a narrow band between Milwaukee and Port Washington. These barns were constructed primarily in the 1890s, many of them by a German carpenter named Ernest Clausing.

Why was there interest in building octagonal barns? One logical reason was that a barn builder in the community knew how to build the eight-sided structure. A farmer with some flair for mathematics could calculate that a round or polygonal structure provided more economical space than a square or rectangular one. Furthermore, the upper parts of a centric barn had clear space, uncluttered with the supporting beams of bank barns. Still others theorized that round or octagonal structures resisted strong winds better than square structures. Some suggested this was the reason a considerable number of octagonal barns were built in Ozaukee County, where strong winds often blow off Lake Michigan.

Centric barns were considered easier to construct than rectangular barns, and cheaper too, because heavy bracing timbers were not required. Historian Richard W. E. Perrin describes the construction of an octagonal barn this way:

> Generally set upon a fieldstone foundation, the superstructure at hay-floor level is one large area. The roof rafters are joined together near the top by an octagonal timber "collar." Similarly, the plates upon which the rafters rest are tied together with angle irons to make a continuous ring of the plate, thus converting the lateral thrusts of the roof into vertical loads upon the outside bearing walls. The roofs of the octagonal barns were generally crowned with a louvered or glazed cupola.[4]

## Round Barns

Writer Eric Sloane says the first round barn of stone in the United States was built by the Shakers at Hancock, Massachusetts, in 1826. This barn is massive—270 feet around and with walls 21 feet high. It housed fifty-two cattle. The center of the Shaker barn was the haymow.

Round barns were once promoted as the most economical and efficient barns that could be built. The University of Illinois conducted extensive research to prove the point. Unfortunately, modern barn technology such as barn cleaners and pipeline milking machines did not work well when used in circular structures. Pierce County, Ellsworth, date unknown. **WHi Image ID 30897**

Sloane claims the early round barns had a religious origin. Such sects as the Quakers, Shakers, and Holy Rollers used the concept of the circle in many religious ways, such as prayer circles. The round barn was seen as an extension of the circle in religious observance; some say these groups built round barns "to keep the devil from hiding in the corners."[5]

During the early 1900s the round barn was promoted as one of the most efficient types of barn to build. Even Sears, Roebuck and Co. offered a round barn plan in one of its barn mail-order catalogs. The chief advantage suggested by its proponents is geometric: a given area is enclosed with a shorter line in the form of a circle than in any other geometric figure. A similar claim was made by those advocating octagonal buildings. Because the round wall can be constructed using the principle of the arch and hoop, a stronger wall can be built, too, often with lighter construction. Circular construction takes advantage of the lineal strength of the lumber, which is some twenty times stronger than its cross-grain strength.

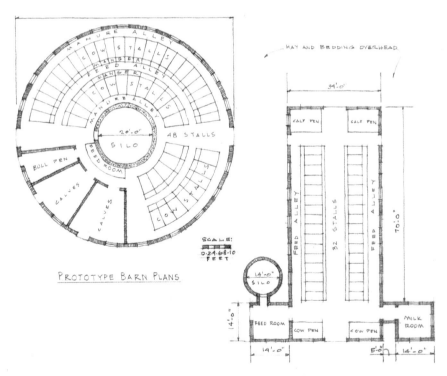

Typical interior arrangements for round and rectangular barns. **Drawing by Allen Strang, courtesy of William Strang**

Another advantage of circular barns over the more common rectangular barns is the need for less framing lumber. No framing lumber larger than 2 inches by 8 inches is required for the upper part of the barn. The circular barn is held together by bands of boards running around the barn, supporting the uprights and tying them together as hoops do a barrel. If the lumber is correctly placed in a round barn, it performs two functions. Every row of siding boards running around the building encloses the building and, at the same time, serves as a brace. And not as much strength is needed in a round barn because the exposed surfaces are curved and resist wind pressure much better than those of a square structure.

The round barn, as well as the polygonal barn, has the main disadvantage that additions to it are difficult. It is relatively easy to add on to a rectangular barn, as has been done with many. But not so with centric barns. Also, having insufficient light in the center of a round barn is a common complaint. Another basic objection to the round barn is that rectangular objects cannot be placed in a circle without wasted space. The assumption made is that cows and horses are rectangular objects, which is incorrect. Cows and horses are wedge shaped, requiring less space in front than at the rear. In round barns, the silo is usually constructed at the center, with the barn built around it. This makes for an efficient feeding arrangement, particularly if the cows are arranged to face the center—as they should be to take advantage of their wedge shape.

The University of Illinois promoted round barns in the early 1900s to a considerable extent. A college bulletin reported on research done comparing the building costs of round and rectangular barns. The researchers sent letters to all the farmers they knew who had constructed round barns, requesting information about the relative costs of building the barn. These are some of the replies they received:

> As to labor, one can build a circular barn much cheaper, because anyone that can use a hammer and saw can work on a circular barn. All heavy timbers, as sills, purline plates, etc., are nailed together one inch at a time.
>
> I built a barn 40 x 60 feet containing 2,400 square feet with 18-foot posts, the carpenter work costing $250. The next year I built a round barn 60 feet in diameter containing 2,826 square feet with 20-foot posts and the carpenter work cost me $240. The round barn contained 15 percent more floor area and the carpenter bill was 4 percent less.[6]

In 1893 J. H. Sanders described a round barn designed by Professor F. H. King of the Wisconsin Agricultural Experiment Station for a dairy farmer near Whitewater.

> The plan of this barn is presented here, not as a novelty in rural architecture . . . but because in several fundamental features it embodies ideas which are believed to be worthy of general imitation, and this particular plan is used simply as a concrete

illustration of the character of improvements which at the present time are greatly needed to insure the highest development of a large and profitable animal husbandry in a climate of long cold winters.[7]

This particular round barn was designed with two stories. A wagon could be driven into the first story through a wide door, into the second story by means of a ramp. The barn was covered with a conical roof topped with a cupola at the peak. A silo, 24 feet outside diameter and 34 feet deep, was built at the center of the barn. Around the silo in two circular rows were stalls for ninety-eight mature cows. The inside row of cattle faced the outside of the barn, the outside row faced inside. Thus the cattle could be fed with a common feed alley, but two wagon drives were necessary for manure removal.

The diameter of the barn was 92 feet, and it was 28 feet from sills to eaves. In the second story of the barn, extending all the way around the silo was a barn floor 18 feet wide. At its outer edges were chutes leading directly to the feeding alley below. Either green fodder from a wagon or dry hay from the haymows could be forked down these chutes to the cattle.

One interesting feature of this barn was the housing for horses—on the second story of the barn, to the right of the entrance, was a stable for ten horses. On the left side of the entrance were a workshop and granary. To the rear of the silo there was a tool room. Between the tool room and the horse barn on one side, and between the granary and the tool room on the other, were two hay bays. Hay storage was also available above the barn floor, granary, and horse barn.

The barn was constructed entirely of wood, on a foundation of four concentric stone walls that extended just far enough above the ground to keep the wood portions of the barn away from moisture. The inner wall supported the walls of the silo. The two middle walls supported the stanchions for the cows as well as the central portion of the floor, main posts, purline plates, and roof. The outer wall supported the wall of the barn.

The frame of the barn consisted almost entirely of two-inch lumber, one of the reasons the costs for construction were kept low. No mortise-and-tenon work was necessary; the construction was done entirely with a hammer and a saw. The outside of the barn was covered with drop siding sprung and nailed to the studding so as to break the joints.

The feeding mangers for the cattle were molded out of earth. A round-bottomed trough was formed in front of each row of cattle. The earth was then plastered with a coat of water lime. The formation of the mangers was interesting because later construction used concrete almost universally for such structures. The description of the barn says nothing about the stable floors themselves.

Claimed to be the largest round barn in the world, this barn is located at the fairgrounds in Marshfield.

The ventilation system was deemed the most modern for its day. The spaces between the studs in the walls of the silo were open from the floor of the cow stable to the cupola on top of the barn. Sanders described the workings of the ventilation system as follows: "The heat given to these flues by the silage in the silo, the warming of the air in the basement by the cattle, and the suction produced by the wind blowing through and around the cupola, all combine to maintain a strong current of air out of the barn through the cupola and in through the gangs of augur holes in the outer walls." Holes were drilled in the outside of the barn wall, near ground level, to admit fresh outside air. As a fringe benefit of this ventilation system, because all of the warm air in the barn escaped up the sides of the silo wall, the wall was kept free of moisture and didn't decay as quickly as it might have if moisture was allowed to accumulate.[8]

With lumber at fifteen dollars per thousand board feet, this barn was constructed for twenty-four hundred dollars. In his summary of the description of this round barn, Sanders wrote: "The great economy of the circular plan for farm buildings over other types of structure diminishes as the size of the building decreases, but it is nevertheless well adapted to some of the smaller structures such as horse barns and sheep barns. In any case where an octagonal barn is desired the circular type will always be found cheaper and more stable."[9]

Though there was some promotion of round and polygonal barns in Wisconsin, and considerable interest in the late 1800s and early 1900s, they did not become a basic building design for the Wisconsin builder. With the exception of those in Ozaukee and Vernon Counties, most centric barns were scattered around the state, probably depending on the particular interest of a farmer or barn builder in a community. For the most part, they were viewed as novelties by other farmers in the community, who depended on their traditional bank barns to serve them.

## ⚜ The Round Barns of Vernon County ⚜

**Vernon County, in the driftless area of southwestern Wisconsin,** boasts one of the largest concentrations of round barns in the country. According to the Vernon County Museum and Historical Society, ten of these original round barns are still standing. The society has prepared a guidebook for those who want to see these interesting old barns on a driving tour.

Alga Shivers, an African American son of a slave from Tennessee, built at least fifteen Vernon and nearby Monroe County round barns. Shivers's father, Thomas Level Ethridge Shivers, was born on a plantation near Alamo, Tennessee. In 1879, with his sister Mary and his brother Ashley, Thomas settled in Vernon County, in what at the time was a large African American farming community. Alga left his father's farm to attend George R. Smith College in Missouri, where he studied engineering and carpentry. After attending college and serving in World War I in France, he returned to

Vernon County boasts the largest concentration of round barns in Wisconsin. This one, in Ontario, Wisconsin, was photographed by celebrated landscape photographer Paul Vanderbilt in 1965. **WHi Image ID 10446**

the home farm and began building round barns, a style of barn he apparently preferred over the more conventional rectangular ones.[1] Shivers followed a building approach that began one to two years before the construction of a barn. He and his crew would cut logs in the farmer's woodlot, and when the wood was cured, Shivers and his crew of two or three would construct the barn. He preferred building his round barns of wood.

**Notes**

1. Peggy Hoehne, "The Round Barns of Western Wisconsin," www.suite101.com/article.cfm/wisconsin/106343/1.

Alga Shivers, a descendant of slaves, built many of the round barns in Vernon County in the years following World War I. Vernon County, County Highway Q one half mile south of Hillsboro.

Though round and polygonal barns, in theory, were the most efficient barns to construct, the Wisconsin farmer thought in terms of squares and rectangles. It's interesting to speculate why this should be. One obvious answer is the system used in surveying the land. The surveyors laid out the land in squares and rectangles. A township was surveyed to be six miles square. A section of land, 640 acres, was a mile square. A 160-acre farm, a common size for many years in Wisconsin, was half a mile on a side.

Because the basic shape of their farms was square, farmers laid out their fields in squares or rectangles. And they took great pride in being able to plow in a straight line, to shock grain in straight lines, and to plant their corn in straight rows. A round barn called for a different way of thinking. Many farmers simply were not ready to accept a round barn in their square world. (Slightly aside, but related to the same point, is the resistance the Soil Conservation Service encountered when it arrived in Wisconsin and promoted plowing around the hills—contour plowing—to prevent soil erosion. After generations of plowing straight furrows, farmers resisted the suggestion that plowing in a curve was a better way.)

To this day Wisconsin farmers see much of their world in straight lines and square and rectangular shapes. It is not surprising that the centric barn did not catch on and become a primary form of barn building in Wisconsin. To cause a change from square thinking to round thinking took more than the sheer logic that showed a round building was more efficient, was easier to build, and resisted wind better than a rectangular structure. Promoters of round barns failed to take into account the need to change basic attitudes toward roundness and straightness.

Several special examples of round and multisided barns can still be found in Wisconsin. Marshfield claims to have the largest round barn in the world. The barn, which was finished in 1916, is 150 feet in diameter and 70 feet high at the highest point and is located on the Central Wisconsin State Fairgrounds on Seventeenth Street in Marshfield. It cost five thousand dollars to build the barn, which is used for cattle exhibits and shows. It can seat from five hundred to one thousand people.[10]

One of Wisconsin's most unusual centric barns—it has twenty sides—is the Nashold Barn in Columbia County, near Fall River. Built in 1911, the barn recently has been restored.

And in far northern Wisconsin, in Iron County, is the Annala Round Barn, constructed entirely of massive fieldstones. It was built in 1917 by a Finnish stonemason, Matt Annala. The site also includes a centric milkhouse with a stone bottle decoration on its roof.

The Marshfield, Nashold, and Annala barns are all listed on the National Register of Historic Places. At the time of this publication, these three special barns still stand. They are worth a trip to see.

Old barn roofs have inherent beauty. Here we see contrasting gable roofs on a barn at Stonefield historic site near Cassville.

# 5

# FORMS AGAINST THE SKY

## *Barn Roofs*

The most obvious characteristic that distinguishes one barn from another is its roof. In Wisconsin three roof types are most prominent: gable, gambrel, and arched. Lesser known but still possible to find are saltbox, hip, snug Dutch, shed, and monitor roofs.

### GABLE

The gable roof is common on many Wisconsin barns as well as other farm buildings, including homes. This type of roof has two planes that meet at the top. The gable roof is found on early log barns, English barns, Norwegian barns, and many of the Finnish barns. The early bank barns—usually those with forebays or pentroofs—had gable roofs. The reason is probably a practical one: they were easier to build than the gambrel roofs that top so many dairy barns today.

The gable-roof style, seen here on both the barn and log hog house, is a common roof style among Wisconsin's old barns (new ones, too). Waukesha County, Old World Wisconsin, Eagle.

Beginning in the late 1800s and continuing for many years, the gambrel-style roof was the most popular for Wisconsin barns. This type of roof provided more storage area for hay than did gable-roof barns. Buffalo County, Highways 35 and 25 north.

## GAMBREL

On the big dairy barns around the state, the gambrel roof is the type most often seen. Though accurately called gambrel, it is also referred to—incorrectly—as a hip roof.

The word *gambrel* comes from the name given the hock, the bent part of a farm animal's back leg, and probably also from the gambrel stick, shaped like a gambrel roof, which was commonly used on most farms to suspend butchered animals.

The gambrel roof has four planes. The two planes at the top have a slight pitch; the side planes are longer and steeper than those at the top. The primary purpose of the gambrel roof was to allow more space for hay storage under the eaves. A person can walk upright under any part of the roof. And the main supports for this type of roof are located near the sides of the eaves, leaving the center area open.

The pitch of both the upper and lower planes of gambrel roofs varies somewhat depending on the width of the barn, its height, and, most likely, the desires of the farmer and the barn builder. Another variance in gambrel

A distinguishing feature of the Dutch gambrel roof style is the flare at the roof's eave line. Pierce County, Highway 35 and County QQ.

This New England–style gambrel-roof bank barn in far northern Wisconsin has a large metal cupola and weather vane. Douglas County, County E and County L.

roofs is the treatment at the eaves. One type of gambrel roof flares out at the eaves; the other continues straight. Historian Eric Sloane calls the gambrel roof that flares out a Dutch gambrel; the one that continues straight, he refers to as a New England gambrel.[1]

Early gambrel-roof barns (those built in the late 1800s and early 1900s) were mainly of post-and-beam construction. Starting as early as the early 1900s, barn builders began adopting newer building approaches that allowed the use of lighter-dimension lumber rather than the heavy posts and beams. Many of these barns, built with so-called plank framing or light framing techniques, and later with balloon framing, had gambrel roofs.

## ARCHED

The arched roof—also referred to as rainbow roof, gothic roof, or round roof—came later, generally just prior to 1940 and extending on into the 1950s. By that time, barn builders had abandoned the more traditional pegged-beam barn in favor of construction approaches using trusses and laminated beams. These newer building techniques allowed barn roofs to be constructed in an arched shape. The arched roof, compared with the gable and gambrel roofs, has maximum storage space for hay. According to H. J. Barre and L. L. Sammet, agricultural engineers, a barn thirty-four feet wide with a gable roof will store 0.7 ton of baled hay per foot of length. A similar barn with a gambrel roof will store 1.6 tons. An arched-roof barn will store 2.0 tons.[2]

Farmers, as they sought to have ever larger spaces for the storage of hay, unencumbered by beams or trusses, looked favorably on the arched-roof barn design. The only problem was, how does one bend a rafter? Early arched rafters were constructed by nailing short pieces of

Arched-roof barns began appearing in the late 1930s, after carpenters began perfecting construction with lighter building materials. Barron County, near Rice Lake Regional Airport, 19th Street and 15th Avenue.

wood that were clamped together and placed in a curved form, a labor-intensive project. The next stage in producing a curved rafter was to bend or spring them together in such a way as to effect a curve. Unfortunately, some of these early arched roof barns, especially those with bent or sprung rafters, were not sturdy and over time sagged, creating a barn with a roof line that bent in the middle like a broken-down workhorse. Some farmers combined both sawed-type rafters with bent rafters to increase the strength of their barn roofs.

By the late 1930s, with the development of modern glues, the lamination of bent-rafter beams took a new turn. The Forest Products Laboratory of the United States Department of Agriculture located in Madison began experimenting with glued-wood products. The laboratory reported that glued-laminated rafters were two to four times stronger than bent and sawed rafters nailed together. Laminated rafters produced and marketed by Rilco Laminated Products of Minnesota soon were widely used and became the gold standard for the construction of arched-roof buildings, including barns.[3]

An arched-roof barn under construction in Viroqua, Wisconsin, circa 1933, shows the details of the elaborate roof trusses and beams. WHi Image ID 24364

The monitor roof design is relatively rare in Wisconsin. Lafayette County, Highway 151 south of Mineral Point.

## OTHER ROOF TYPES

The occasional hip-roof barn is seen in Wisconsin. A hip roof is one that is pitched in all four directions. While some Wisconsin farmers refer to the gambrel-roof barn as having a hip roof, this is incorrect. The hip roof is more difficult to construct than the gable roof and has no particular advantage over other roof types. It seems to be the personal preference of the builder that dictated whether a barn should have a hip roof.

The snug Dutch roof, also known as a clipped gable, is related to the hip roof. To picture the snug Dutch, think of a gable roof that is pushed back from the gable end. The tops of the gable ends are flattened, but not to the extent of a hip roof. There is no particular advantage to the snug Dutch roof, though it does set off a barn from others in the community.

Some barns have monitor roofs, in which the center part of the structure has a gable roof and attached below the eaves on either side are shed roofs, forming the monitor roofline. Many barns, especially in southwestern Wisconsin, are basically gable-roof barns with a long sweeping side, giving them a somewhat monitor-roof appearance.

This brings up the interesting instances of combinations of roof types formed when additions were made to barns. The early dairy barns were large and spacious for their time. But as dairy farms grew in acres and as more cows were added to the herds, the barns also grew. Rather than tear down an old barn and start over, which seems the approach used in much urban construction, the dairy farmer figured out ways of adding to the barn to give more space for livestock and more room to store feed.

The saltbox roof resulted when a section was added to a gable-roof barn, extending the roof line on one side until it came within a few feet of the ground. Though additions were often responsible for the saltbox shape, some farm buildings were originally built with this roof type. If possible, the long side of the roof—the side that came closest to the ground at eaves edge—was placed toward the north as a protection against the winter wind.

Both gambrel-roof and gable-roof barns in Wisconsin have a variety of additions. A common one is a shed-roof lean-to attached to one or both ends of the barn, which was often an open area used as a shelter for young stock. If the farmer raised a variety of livestock, it was a shelter for beef cattle or hogs.

More often than adding a lean-to, the dairy barn was lengthened to accommodate more cattle and more feed. Sometimes the roof line was simply extended, and after a few years the change was hardly noticeable. But most of the time the addition, either to the length of the barn or in the form of an L, carried a roof type different from the original barn. The combinations varied. It's not difficult to find gable-roof barns with arched-roof additions,

This addition appears to be a storage shed, or perhaps a place to house young stock. Note the tile silo. Richland County, County Road E and Basswood Road, about six miles south of State Highway 171.

gambrel-roof barns with arched-roof additions, and even gable-roof barns with gambrel-roof additions. The roof type tells something about the time the addition was made. If the roof type was arched, for example, it probably was made after World War II. It is not uncommon to find dairy barns with two or three additions, each with a different roof type, lending a rather cluttered appearance.

## ⊰ Today's Dairy Barns ⊱

**Today's modern dairy barn is one story,** often constructed with steel framing and a gable roof, fifty or sixty feet wide and often several hundred feet long. These barns have canvas covers on the sides that can be lifted at the ends to provide ventilation. Usually bunker silos (horizontal with concrete sides, an open top, and one open end) or a series of plastic tube silos or both are located nearby. Sometimes several of these barns are clustered together, each housing several hundred milk cows. These barns are manifestations of the large dairies that have emerged during the late 1900s into the first decade of the twenty-first century, referred to by some as factory farms or "confined animal feeding operations" (CAFOs). These operations, with their modern buildings, produce enormous quantities of Wisconsin's milk.

This modern, metal dairy barn boasts five silos. Dane County, Schumacher Road off Highway 113.

For large-scale dairy farmers, the old two-story wooden barns do not fit modern-day dairy operations. Some of these farmers have discovered ways of reusing the old barns for housing heifers, storing machinery, or retrofitting into milking parlors. Others have torn down the old barns to take them off the farms' tax rolls.

One of the most rapidly growing groups of dairy farmers in Wisconsin farm organically and pasture-graze their cattle. (In 2009 Wisconsin was ranked number one in the nation in organic dairy production.) These farmers tend to have much smaller herds—often fewer than seventy-five cows—and many of them use existing old wooden dairy barns for their operations.

Graziers are dairy farmers who follow principles of rotational grazing, meaning that during the summer months their cattle are out on pasture and not confined in a barn. (Some graziers, but not all of them, also farm organically.) In 2004 approximately 14 percent of Wisconsin's dairy farmers (about 2,100 farmers) grazed their cattle. Fifty-eight percent of Wisconsin's graziers had dairy herds of fifty cows or fewer, and most of them are using original dairy barns, sometimes with additions and often with modifications.[1]

**Notes**
1. Wisconsin Agricultural Statistics Service, "Wisconsin Dairy Grazing Operations 2004 and 2009," Madison: July 2005.

Often, several roof styles collided as farmers built additions to their barns as they expanded their dairy herds. Here an arched-roof addition has been attached to the original gambrel-roof barn. Chippewa County, County Road OO west of 180th Street southeast of Chippewa Falls.

A variety of barn roof types survive around the state. The early barns were built to last, and many farmers, being the practical economists they are, will continue to use them for years to come. Though they may add to the barns as the size of their operation increases, or even add new, more modern barns nearby, the original barns will stand as long as the farmers can find uses for them. Economics often determines whether an old barn will continue standing or be torn down.

## Roofing Materials

Some of the earliest barns in Wisconsin had thatched roofs. Thatching was common in parts of Europe, and many immigrants brought thatching skills to this country. The Finnish are reported to have roofed some of their barns with wide boards, similar to what might be put on the sides of a barn, and some early barns were shingled with bark, though the irregularity of bark shingles made them extremely difficult to work with.

Wood shingles soon became the most popular roof covering for barns and continued to be used well into the 1900s. Early wood shingles were cut by hand with a tool called a froe. The froe was struck with a wooden mallet to slice shingles off a block of wood. Cedar was the most popular wood for shingles due to its natural weather resistance. However, by the time most Wisconsin dairy barns were built, wood shingles were made commercially.

A characteristic of wood shingles, a sort of fringe benefit when they were applied to a roof, was their ventilation quality. Standing inside a wood-shingled barn on a sunny day in summer after several weeks of rainless weather, the farmer could see sunlight streaming through his barn roof in a thousand places. But as soon as it started to rain, the wood shingles swelled, and the roof did not leak at all.

This expanding and contracting of the wooden shingles tended to loosen the nails holding the shingles. An old wood-shingled roof resembled a pincushion, with hundreds of nail heads protruding. Also, on the north side of a wood-shingled roof, it was not uncommon to find moss growing. This seemed to cause no particular difficulty for the roof, and it even gave the roof a bit of character. Some believed the

Barn builders commonly used cedar shingles to roof barns, especially before asphalt and other roofing materials became widely available. Waukesha County, 1870s horse barn at Old World Wisconsin, Eagle.

moss rotted the shingles, but I know of a mossy wood-shingled roof that is sixty-five years old and leaks not a drop.

For a time, metal-roof barns became popular in Wisconsin. Some farmers even roofed their barns with rolled roofing paper. The problem with both metal and rolled roofing was that a strong wind could tear it loose. In addition, the metal roofs, particularly the earlier ones, rusted rather easily.

Beginning in the 1920s, many barn roofs were shingled with composition shingles of the same type put on houses, a method still in common use today. But underneath many composition roofs, the barn detective will find cedar shingles. Now the farmer standing inside his barn on a bright sunny day sees no sunlight seeping through the roof. If he does spot a hole, he knows Junior has been practicing with his .22.

## ⚡ A Field Guide to Barn Roofs ⚡

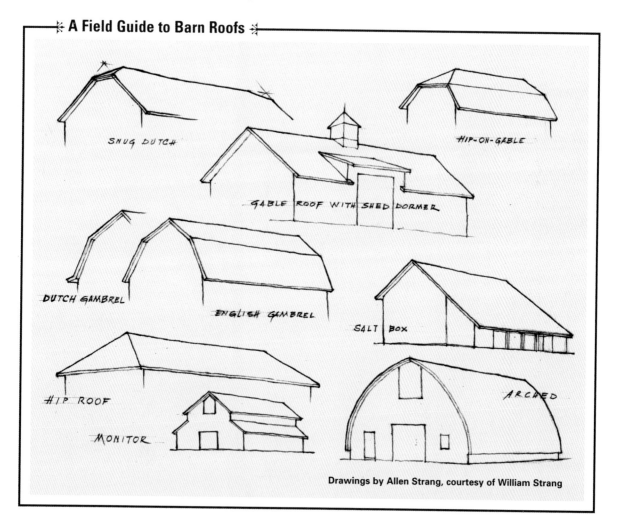

SNUG DUTCH

HIP-ON-GABLE

GABLE ROOF WITH SHED DORMER

DUTCH GAMBREL

ENGLISH GAMBREL

SALT BOX

HIP ROOF

MONITOR

ARCHED

**Drawings by Allen Strang, courtesy of William Strang**

Barn walls are beautiful creations, beyond the practical. The shape and color of stone, mortar, and a wooden door add to the artistry. Grant County, Stonefield historic site, near Cassville.

# 6

## Stone, Sawdust, and Sweat
### *Builders and Building Materials*

Understanding building materials—which type of wood should be used for the beams, which for the flooring, which for the boards that would face the weather—was a basic skill of the barn builder. He knew about wood grain and tensile strength, about snow loads, about how much bracing was required so a barn would withstand the storms so common in the Midwest.

In addition, the barn builder had a deep love for the materials he used. He enjoyed the smell of freshly cut pine boards and cedar shingles, and the feel of planed lumber. The barn builder also possessed a sense of wholeness about building, an insight that allowed him to envision the completed structure and how each operation, each board, each timber, every detail, contributed to that whole. Thus, a barn was not just a series of timbers pegged together, with so many hundred boards nailed to the outside. It became an entity that was more than the sum of its parts. The early barn builders were craftsmen, in the best sense of the word.

Except for pioneer log structures, farmers usually did not build their own barns. Nearly every community had at least one carpenter who was a barn builder; many communities had more than one. My home community in central Wisconsin had two barn builders living within a half mile of each other, both of Welsh extraction. Though they didn't work together, they each built the same basic type of barn—a gambrel-roof bank barn. They also both built farmhouses. One became quite successful at house building; the other received complaints from his customers. It seems the second carpenter had grown so accustomed to working with huge timbers and framing the big barns, he couldn't shift his thinking to such small-scale projects as constructing cupboards and shelves. He did well on the framing part of the operation, however.

The mainstay of the old barns was their framing timbers. Like the bones of a human, the timbers formed the skeleton of the barn. The timber frame gave the barn its strength and form, outlined its size, and even determined which parts of the barn would be used for what purposes.

Until sawmills came into the country, timbers were hewed with a broadax. Of course, sawmills followed close behind the settlers in Wisconsin and were often built beside a fast-moving stream so the mill could be powered with a waterwheel. As a result, hewed barn timbers are somewhat difficult to find. Hewing timbers was a skill passed on from craftsman to craftsman over the years. The trees were first cut with a felling ax. Then the limbs and the bark were removed. The broadax is often confused with the felling ax. Some people, upon first seeing a broadax, mistakenly believe it was used for felling trees. This is entirely wrong, for the huge broadax, with a bit often as wide as fourteen inches and weighing up to nine pounds, was far too clumsy for felling.

At a barn raising in the town of Deerfield (date unknown), crew members mortise freshly hewn timbers with a hand-operated boring machine. **WHi Image ID 26256**

After the log was prepared, the hewer and his helper snapped a chalk line—a string rubbed with chalk—along the log, marking a line the length of the log. With a long-handled scoring ax, which had a narrow blade and a thick bit, the skilled craftsman made a series of cuts about six inches apart along the length of the log. Then the hewer took the broadax and came down the scored line at an angle that would slice a thin slab off the side of the log. The broadax was beveled on only one side, and the handle was often curved so the worker's hands wouldn't hit the log while he worked. There were both right-handed and left-handed broadaxes. The broadax removed the wood between the cuts made with the scoring ax, leaving the entire side of the log flat. The procedure was repeated until the log was flat on all four sides, and thus square. The hewed timbers—6, 8, 12, or more inches square—ranged to forty or more feet in length and were usually hewed at the building site.

The view from the threshing floor to the ceiling of a post-and-beam-constructed barn shows the intricate craftsmanship used for tying posts, beams, and braces together. Waukesha County, German barn at Old World Wisconsin, Eagle.

By the time that most of the big dairy barns in Wisconsin were constructed, in the late 1800s and early 1900s, many sawmills operated in Wisconsin, and the barn builder could buy ready-cut timbers. Even with the advent of sawmills, many farmers used floor joists and roof rafters of unsawed material.

Often a farmer, particularly one who had access to a tamarack swamp, cut tamarack poles six or more inches in diameter and used these for floor joists in his barn. Many of these farmers didn't even remove the bark from the tamarack poles, leaving them in their natural state. One can find such floor joists intact in some old Wisconsin barns, covered with years of whitewash applications. Another common practice was for the farmer to flatten one side of the poles used for floor joists, the side to which the mow floor was attached. This made nailing down the mow floor easier. The support posts in big dairy barn basements were often huge posts left in their natural state. Usually the bark was removed, sometimes not. Same for the barn's rafters. One could often find tamarack poles used as rafters, hewn flat on one side so the roof boards could easily be attached.

These were obviously ways the farmer believed he could save money in building his barn. He didn't have to purchase all of the timber or even take his own logs to the sawmill for cutting. Many farmers did have their own logs sawed, however—for timbers, for lighter framing wood such as rafters and floor joists, and for the barn boards that covered the outer surface of the building. When this practice was followed, the farmer had to plan ahead so the wood had an opportunity to age before he started building his barn. Green wood often split, warped, and shrank when it dried out, leaving the farmer who had built with it very disappointed indeed.

The haymow floor of an old barn is likely to have cracks between the boards. Some argue that the cracks were left there on purpose, for additional ventilation. More accurately though,

In most of the glaciated areas of Wisconsin, fieldstones were readily available and were used by early barn builders to construct barn walls. Fieldstone walls were both sturdy and attractive. East Marathon County, northwest of Town Line Road.

in a natural aging process, the wood gradually shrank over the years, pulling the boards apart and leaving the cracks. When most of these barns were built, the boards fitted together tightly. In some barns, the width of cracks in the haymow floor became rather annoying. Hay leaves and dust sifted through the cracks, spilling down the necks of those caring for the cattle and too often finding their way into the fresh milk.

In some of the many different types of barns constructed in Wisconsin, wood reached from the peak of the roof to the ground, supported by a small masonry wall to keep the wood from coming into contact with the soil and rotting. Many barns were built on masonry basements, particularly in parts of the state where masonry building materials were handy and inexpensive. One of the common types of barn basement wall was constructed of fieldstone.

Just as the early farmers in Wisconsin had a ready supply of trees to give them building materials for their barns, many also had a ready supply of fieldstones. Both the trees and the stones had to be removed before farming could take place; their use as building materials was

Welsh barn builders sometimes built barns entirely of rocks and mortar. This gable-roof bank barn is an example. Columbia County, Highway 22, one mile north of Highway 33.

a bonus for the farmer. In the unglaciated or "driftless" area, limestone lay close to the surface and was commonly quarried for foundation material.

Much of Wisconsin's landscape is a product of the Ice Age; the last glacier melted and receded some ten thousand years ago. As the glacier pushed south, it brought with it stone fragments and boulders torn loose from the bedrock hundreds of miles to the north. Only the hardest stone—granite, basalt, and quartzite—could withstand the grinding action of the glacier without being crushed into sand and gravel. For this reason, when sandstone is found in the glaciated areas of the state it has seldom been carried far from where it originated.

The boulders from the north were mixed with the soil, in some parts of Wisconsin so densely it was nearly impossible to remove enough of them for farming. Many fields in the state have been cleared of trees but not of stones and boulders. It would appear that the farmer, once he saw how many stones there were, decided to leave the field for pasture.

Another problem that farmers of stony soil have is that the stones move to the surface with the freezing and thawing of the soil each year. Thus a farmer may have picked his field clear of stones one spring, only to find that the job has to be repeated the following spring before the crops can be planted.

**Some early silos were built of fieldstone. Waukesha County, Highway 18 and County G.**

In some parts of Wisconsin, especially Door County, one can find barn walls sided with wooden shingles. County EE and County F, Bailey's Harbor.

Just as there were carpenters skilled in barn building in nearly every community, so also were there skilled stonemasons who could take the raw stones from the fields and form them into walls that were both attractive and extremely sturdy. Fieldstones were prevalent in much of northern Europe and had been used there as building materials for centuries. Many stonemasons had immigrated to Wisconsin, and these masons found fieldstone readily available in many parts of the state.

Fieldstone walls were most popular in Wisconsin between 1840 and 1880 and remained popular in some parts of Wisconsin until as late as 1920. After 1900 fieldstones were used primarily for barn walls and foundations, for silos, and for smaller buildings such as pump houses and smokehouses. In addition, a few barns were built entirely of fieldstones.

Some stonemasons became quite expert at cutting and fitting fieldstones together in a wall, using a minimum of mortar to hold the stones together. A few such walls can be found in Waukesha County. The next evolutionary step was for the stonemasons to square off the stones and lay them as they would lay building blocks. Some barns had squared fieldstones at the corners and around the doors and windows but used round ones for the rest of the wall.

In addition to fieldstones, quarried rock was used in the building of barn basements and, in a few instances, entire barns. Particularly in southwestern and western Wisconsin—a part of the state the last glacier missed—it is common to find barn walls constructed of squared quarried rock.

In a few instances bricks were used in the construction of barns. As Fred Holmes noted in *Old World Wisconsin: Around Europe in the Badger State,* the Belgians who settled in southern Door County built wooden structures. When the great Peshtigo Fire of October 8, 1871, swept through their area, they lost all their buildings. "Required to start anew, the inhabitants dug the red clay soil, moulded it into forms, dried it in the sun, and then kiln-burned it."[1] Bricks were also used as nogging in the German half-timber buildings, and stovewood was used both for nogging in half-timber construction and for free-standing barn walls.

Beginning about the turn of the twentieth century and continuing into the 1900s, concrete became popular for barn construction, used for foundations, floors, mangers, and even the walls of barn basements. Concrete blocks also gained favor in the early 1900s. In the early days, some masons made the concrete blocks right at the building site, although by the early 1900s factory-made concrete blocks were commonly available. In its 1908 catalog, Sears, Roebuck listed a Wizard Concrete Block Machine that sold for $42.50. It made concrete blocks 8 inches by 8 inches by 16 inches. Sears described the Wizard machine as "the best concrete building block machine made. It contains all up to date scientific improvements known to this class of machinery. In offering you this high grade machine, we claim it has no equal and that it is superior to concrete block machines which are sold as high as from $100.00 to $200.00."[2]

By the late 1930s concrete blocks began replacing fieldstone in the construction of barn walls. Concrete blocks were easier to work with, as sturdy as fieldstones, and readily available in most communities. This L-shaped barn has concrete block stable walls. Washburn County, County B north of Highway 53, one half mile north of Sarona.

Building with concrete was not a new idea, although concrete was not used in this country much before the middle 1800s. Joseph Goodrich is credited with having built the first poured-concrete building in the United States, at Milton, in 1844.[3] It was noted both for its unusual hexagonal design and for the material used in constructing it. (Milton House has been preserved as a museum and is open to the public during the summer months.)

The use of cement in building goes back four thousand years to the Egyptians, who knew how to make a cement that hardened in the presence of water.[4] The Romans, in the construction of sewers, aqueducts, buildings, and roads, used cement of such good quality that specimens have survived to this day. Some of the prehistoric people of America—the Aztecs and Toltecs— used a cement mortar so strong that the masonry joints used in constructing stone walls remain after the stones have weathered away. For some reason, however, during the Middle Ages people quit using cement and turned to lime-and-silt mortar for building construction. This material turned out to be far less durable than concrete.

An English engineer, John Smeaton, is credited with rediscovering a method of manufacturing cements from natural rock. In 1756 he found that argillaceous limestones, those containing clays, produced cement that would set when mixed with water.[5] The natural cements were made by heating the limestone to 2,500 degrees Fahrenheit (F) and then pulverizing it. However, it wasn't until after the Civil War that there was much call for natural cements. Then, because of the demand for reconstruction, their use increased considerably.

Another Englishman, Joseph Aspdin, invented artificial cement in 1824. He called his product Portland cement because, when hardened, it closely resembled the natural stone quarried on the Isle of Portland, a peninsula on the south coast of Great Britain. David O. Saylor of Coplay, Pennsylvania, is credited with first manufacturing artificial Portland cement in the United States, in 1871. It quickly surpassed the natural cements in demand, primarily because it was consistent from bag to bag; natural cements often varied considerably in ingredients. Portland cement contains about 60 percent lime, 10 percent alumina, and 25 percent silica, plus lesser amounts of iron oxide and gypsum. It is manufactured by grinding and mixing together these materials, heating them to 2,600 degrees F or 3,000 degrees F, and then grinding the resulting clinker to a powder.

Craftsmen using Portland cement in barn construction quickly learned that adding rocks would strengthen the concrete product. Such metal items as bedsprings, discarded wire, and parts from junked farm machinery were often added to the concrete mixture to give it more strength.

There was considerable discussion among farmers about the virtues of poured-concrete barn walls versus those constructed of concrete blocks. In most instances concrete blocks won the day. Farmers were able to buy them locally, and they were considerably easier to work with than the elaborate forms necessary with poured concrete. Many stonemasons became quite adept at constructing forms for the pouring of concrete, however, and learned how to smooth and shape the concrete to meet the demands of the barn builder. Starting in the early 1900s, poured-concrete silos also became popular. Poured concrete was used almost universally in construction of foundations for barn walls and of stable floors. It hardened to a smooth surface, if the craftsman was at all skilled, and it was extremely tough and long lived. For barn foundations, no forms were necessary for the parts that were underground. The earth sides served as a form for the concrete.

By the early 1900s, because of timber scarcity and because many people couldn't afford to build post-and-timber barns, farmers turned to other barn-building methods. This was the beginning of the light construction building era for barns, a trend that continued into the 1950s. Rather than posts and timbers, barn builders used dimension plank framing, which was cheaper than timbers and more easily managed. (Dimension lumber is that which is cut to standardized thicknesses, widths, and lengths—2 inches by 4 inches by 10 feet, 2 inches by 8 inches by 12 feet, and so on—at a sawmill.)

By this time, farmers were also demanding more open spaces in the haylofts of their barns, and the plank-frame barn builders answered the call. With post-and-beam construction, huge beams interrupted the haymow space at regular intervals, depending on the length of the barn. But by employing techniques used in bridge construction, these plank-frame builders made trusses that created a vast open space in the haymows.

Inside a Sears, Roebuck barn the light construction approach is evident. Sauk County, west of Baraboo.

One of the great attractions of old wooden barns is the wood itself. Whether painted or allowed to weather naturally, wood presents an array of attractive colors and for many people evokes memories of the land. Crawford County, County Road X and Adams Road.

John L. Shawver of Ohio developed the Shawver Truss, which was heavily promoted in farm publications and became quite popular among farmers. The Shawver Truss was a transition phase from post-and-timber to even lighter-weight building approaches, such as balloon-frame construction. Following the balloon construction approach, each rafter and stud shared some of the roof load, and the building did not depend on elaborate trusses to provide stability to the structure. Balloon construction was even less expensive than plank-frame barns. By using light studs (two-by-fours) and horizontal siding, the technique distributed the stresses brought on by wind, roof load, and the barn's contents throughout its framework.[6]

About the time of World War II, steel began to be used widely in barn construction, replacing wooden timbers and posts, wooden roofing shingles, and eventually wooden siding. Many lamented the passing of wood as the primary building material for barns. Wood evoked special feelings—of warmth, richness, and beauty. The natural beauty of wood, whether painted or weathered, was unmatched by other materials. For many people wood was a link to the natural world, where steel and plastic seemed intrusions.

Wooden barns created feelings of oneness with nature, complementing the settings in which they were placed. With the passing of the wooden barn, one more human tie with the natural world was severed. But the economics of barn building and of agriculture forced the shift from wood to steel and other buildings materials. Few people gave any thought to what this change in building materials ultimately would mean.

## ⊰ Sears Barns ⊱

**For anyone older than fifty,** Sears, Roebuck and Co. and its catalog provide fond memories. Richard Warren Sears, a young railroad telegrapher, started the company in 1886. By the early 1900s, after Sears moved to Chicago and teamed with Alvah Roebuck, the Sears, Roebuck spring and fall catalogs found their way into nearly every rural mailbox in the country. These catalogs offered practically everything a person needed, from clothing and farm equipment to patent medicines and musical instruments. In 1908 Sears began offering house plans, and by 1915 one could order an entire precut house through the catalog. Sears sold several thousand homes this way until 1940, when the Great Depression took its toll on sales.[1]

In 1911 Sears added barn plans to its catalog, followed in 1915 by plans for outbuildings such as hog houses, corncribs, and chicken coops. From 1918 to 1930 Sears offered an all-barn catalog called *The Book of Barns.* According

Starting in 1911 and continuing to 1930, a farmer could order a barn from a Sears, Roebuck catalog. The barns were precut and arrived ready to be assembled. Sauk County, west of Baraboo.

to authors Rebecca Hunter and Dale Wolicki, Sears offered seven basic barn styles: gambrel-roof barns (nineteen models), gable-roof barns (seven models), rounded-roof barns (five models), and single examples of an octagonal barn, a round barn, a "clipped"-gable model, and a cross-gable model.[2]

The Book of Barns, Sears, Roebuck and Company, 1922

Starting in 1918 the Sears barns were precut; essentially they were barn kits ready to be assembled on-site. A purchaser had many choices. For instance, the gambrel-roof "Hiawatha Bank Barn No. 3023" could be ordered in any of forty-seven sizes, from 30 feet wide by 32 feet long to 40 feet wide by 98 feet long. Prices ranged from $1,316 for the smaller barn to $3,807 for the larger one. Sears promised its barn customers the following: all siding of cypress, roof boards, flooring, subflooring, hardware, paint (enough for two coats), roofing guaranteed for fifteen years, free building plans, wall plates, girders, joists, studding, ridge boards, plates, rafters, trusses, girts, all posts, all doors, all windows with frames ready cut.[3]

After ordering a new Sears barn, the farmer waited for the local depot agent to inform him of its arrival by railroad. He would then haul the precut pieces to his farm and likely enlist one or more carpenters to assist in the construction, including building a basement for the barn or putting in footings if the barn was not a bank-barn style.

Sears barns can still be found in Wisconsin, but with some difficulty, as many of them appear—at least from the outside—similar to the thousands of other barns that were being erected at the time using more conventional building approaches.

**Sears barn kits came complete with boards, nails, doors and windows, and an instruction book—everything needed to construct the barn except the tools and the hard work.** *The Book of Barns,* Sears, Roebuck and Company, 1922

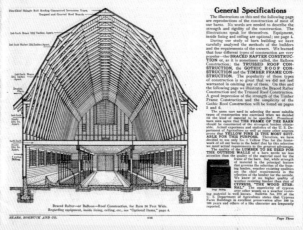

**Notes**

1. Rebecca Hunter and Dale Wolicki, *Sears, Roebuck Book of Barns: A Reprint of the 1919 Catalog* (Elgin, IL: Rebecca Hunter, 2005).
2. Ibid., iii.
3. Ibid., 16–17.

Building a barn required the skills of a craftsperson, the eye of an artist, the patience of a teacher, and the strength of an ox. The task was not easy in the days before electricity, hydraulic lifts, and other labor-saving innovations. Yet thousands of them, like this one, were built in Wisconsin starting in the late 1800s. Jefferson County, Highway 106, seven miles east of Highway 73.

# 7

# WITH THESE HANDS

## *Building the Big Timber-Frame Barns*

From the midnineteenth to the early twentieth century, building the big dairy barns required craftsmanship, hard work, and no small amount of time. A farmer usually planned on a good part of the summer to put up a barn, although he hoped the barn would be sufficiently completed by July so he could fill the mows with hay. Whether or not his hope came true depended on many things—the size of the planned barn, the size and ability of the building crew, and, of course, the weather.

Most of the big barns in Wisconsin were constructed in two stages. The masonry wall for the first story, or basement, was put up, and then the second story, the wooden part of the barn, was erected.

## BUILDING A FIELDSTONE WALL

The first step in building a fieldstone wall was to haul wagonloads of stones to the building site. As many as one hundred loads of fieldstones might be required in constructing a barn wall 34 feet wide by 48 feet long by 8 feet tall. For farmers in the glaciated areas of the state, the stones were readily available. They had been hauling them off their fields each spring ever since they owned their farms. The stones of various sizes, ranging from fist-size stones to those weighing a couple of hundred pounds, were piled near the barn site.

The action of the glacier left many different kinds of stones in the farmer's fields. All were potential candidates for the barn wall, except those some stonemasons called blue

Stonemasons, using the fieldstones at hand, carefully fitted the stones together with mortar to build the walls for the old barns. Besides supporting the barn, the result was a work of art. Waukesha County, Old World Wisconsin, Eagle.

stones. These were avoided because they were hard and impossible to split with any precision. Otherwise, neither shape nor color mattered. Some of the stones were nearly round, others nearly rectangular, but most were variously shaped and colored.

The crew building the wall usually consisted of three men: the stonemason and two helpers. One helper handed the stones to the stonemason; the other helped mix and haul mortar. After they determined the exact location of the building and figured out the location of the walls, they dug a trench about a foot deep and about two feet wide for the wall. Seldom were the walls narrower than two feet, though occasionally they were three or more feet wide. The narrower the wall, the smaller the stones had to be or the larger the number that had to be split. The larger and more massive the wall, the stronger it would be. Both of these factors influenced the width of the barn wall. The preference of the stonemason also played a part in the decision making. Many stonemasons had learned how to build one type of wall. The size of the wall the mason was most comfortable building usually became the size of the wall he built.

To start building the wall, the crew dumped mortar into the trench, followed by stones of various sizes. Because this part of the wall was underground, they took little care as to the

With the fieldstone wall constructed, a barn-raising crew assembled the posts and timbers to create the basic structure of the building. **WHi Image ID 1980**

shapes, colors, and smoothness of the stones. The important consideration was mixing the stones and mortar such that mortar came into contact with all the stones to prevent the wall from shifting once the weight of the rest of the wall was placed on top.

After completing the underground portion of the wall, the crew began the aboveground part, usually constructing about two feet at a time. In a bank barn, where a portion of the wall was built against a hillside, the side that faced the dirt received little attention. This type of wall was referred to as a single-faced wall because only that part on the inside of the basement was made smooth. A double-faced wall was one in which both sides were smooth, as in a non-bank barn and in the walls of a bank barn located on the side opposite the bank.

Using a four-foot-long level and a plumb bob to keep the wall straight, the stonemason slowly built up the wall. No forms were used, except around the doors and windows and, to an extent, at the corners. As the wall got higher, scaffolding was constructed so the stonemason could fit the stones easily into place. Because the sizes of the stones varied so much, the stonemason used his experience and intuition to determine which stone should be placed where. Small stones were usually interspersed with larger stones. And the larger stones were usually faced—that is, a side was split off them so they fit into the wall and at the same time left the wall smooth.

The stonemason used hammers of various sizes to split the rocks. The heavy hammers weighed from eight to sixteen pounds; the lighter hammers were available in weights of three and a half, four, and four and a half pounds. Incidentally, the 1908 Sears, Roebuck catalog offered a sixteen-pound stonemason's hammer, without handle, for ninety-five cents. A three-and-a-half-pound stonemason's hammer was offered at forty-four cents.

The skilled stonemason looked at a stone and determined the direction of its grain, much as a wood splitter determines the grain of a block of wood. Then the stonemason struck the rock, splitting it exactly where he wanted. The stonemason did his own rock splitting most of the time, not relying on his helpers to do the task.

At the corners of the building, boards were set as high as the wall was planned. A string was run from corner to corner as a guide for the stonemason as he worked his way along the wall. The corners themselves presented quite a problem. Imagine the thickness of a corner when the wall itself was three feet thick. Considerable skill was required to make the corner straight; often stones had to be faced on two sides. Occasionally special materials were used in the corners, such as squared stones or concrete blocks, but often the stonemason relied on the materials at hand.

The larger stones were placed at the bottom of the wall, and the smaller ones went on top. This made sense because all of the stones were lifted by hand. For the windows and doors, a

In Wisconsin's glaciated areas, many barn builders used fieldstones to construct barn walls. The fieldstones came in a variety of colors, adding beauty to the barn. Wood County, County Road B, nine miles south of Marshfield.

wooden frame was constructed, and the rocks were built around the frame. For a wall 24 inches thick, two 12-inch planks were put together to make the frame. But the doors and windows were planned so they extended to the top of the wall. Thus it wasn't necessary to pile rocks on top of the window and door frames. In those days, steel lintels weren't available for spanning the tops of windows and doors; they would've been necessary to support the weight of the stones.

A stonemason with two helpers and a ready supply of fieldstones could build a double-faced wall 34 feet by 48 feet in six to eight weeks. Of course the time necessary was affected by the weather, the skill of the stonemason, and the quality of the helpers. Speed, however, was not of primary concern when building a fieldstone wall. Attention to detail, careful measuring, cutting rocks to fit just so, and the proper use of mortar made for a strong, straight wall.

It's doubtful that most stonemasons spent much time thinking about which color stone should go where. They were more concerned with which size stone *fit* where. Nevertheless, the end product, the completed fieldstone wall, was almost always exceedingly attractive. The colored stones—blacks, reds, browns, and grays—the contrasting grayish white mortar, and the informal collection of shapes combined to provide a most pleasing appearance. The massive nature of the walls compared with the subtle colors made a nice contrast, too. When the barn was erected on top of the wall, the picture was completed, for the straight lines of the barn boards, and the red color they were usually painted, provided another interesting contrast with the shapes and colors of the fieldstone wall. The massiveness of the wall demanded that a massive structure be placed on top. Compared with the home and other buildings in the farmstead, most barns were indeed huge structures.

After the first decade or two of the 1900s, few fieldstone walls were constructed in Wisconsin. It became more and more difficult to find a skilled stonemason. But a more important reason was that concrete-block or poured-concrete walls could be constructed much more easily, quickly, and inexpensively. Also, with the use of lighter barn frames, the massive walls required for timber frame construction was no longer required. These walls were less massive, often no more than a foot or so thick, but their attractiveness didn't begin to match that of a fieldstone wall. Thus another era in barn building passed. Today, some renewed interest has occurred in fieldstone construction, but not for barn walls. Some homeowners occasionally add fieldstones as a portion of a house wall, but this is often only a veneer of thinly cut stones that serves no function other than beauty. Today much of the decorative stonework on homes and office buildings is simulated, made from various types of plastics. The day of the fieldstone wall that is two to three feet thick has passed.

Gable-roof barn with a concrete block wall. Concrete blocks had essentially replaced fieldstones for barn wall construction by World War II. Pepin County, County J west of Pine Creek Road.

Still, traveling around Wisconsin, particularly in those parts of the state where the glacier deposited many fieldstones, one can easily find many very beautiful and still functional barn walls made of this natural material. Central and northeastern Wisconsin are good places to look for fieldstone barn walls.

It is also common to find fieldstone walls standing by themselves, their barns either burned or torn down. The heat of the burning barn seems not to have adversely affected the walls. They stand as lonesome reminders of what once was. Standing inside such an old barn basement, with the top open to the elements, the windows long since broken, the doors

As this old barn deteriorated, the fieldstone wall remained straight and true. Waushara County, County O and Highway 22.

rotted and gone, one can't help but think about the stonemason and his helpers, slowly placing rock against rock, filling the open spaces between them with liberal amounts of mortar. One can picture the stonemason's helpers mixing mortar in a mortar box with a hoe and carrying buckets of mortar to the stonemason, who worked hunched over the wall, carefully fitting and measuring.

Here, the future was being laid a stone at a time, yet too often, because of a natural disaster, economics, illness, or a host of other reasons, the old fieldstone barn basement stands alone, the last remnant of what may have been a prosperous farm.

## Erecting the Barn

Once the wall of the barn was completed, the stonemason and his crew departed, and the carpenters arrived. The head carpenter on a barn-building project usually had five or six assistants.

On top of the wall, the carpenters laid a huge beam, the sill, all the way around the barn. This beam was usually 10 inches by 10 inches, occasionally larger. The sill was mortared to the top of the wall to make sure no air spaces remained between the wood and the wall. When the sill was in place, the carpenter proceeded to cut notches in it for the floor joists and for a beam that would run through the center of the barn, from the top of the wall on one end of the barn to the top of the wall on the other end. Floor joists were usually 3-inch by 8-inch beams set 24 inches apart. The floor of the barn had to be particularly well supported, for on it rested the haymows.

Once the sills and crossbeam were in place, the carpenter and his helpers laid the floor of the barn. There were two theories about how the floorboards should be placed. Some suggested they should be a quarter inch or so apart, allowing an air space to ventilate the stable below. This approach caused two basic problems. First, the hay often spoiled just above these cracks in the floor, for all of the moisture coming from below caused rotting. Second, hay leaves and dust constantly filtered through the cracks. The second approach was to use matched lumber for the floorboards so the floor would be tight, allowing no moisture to pass through from below and no dust and dirt to pass through from above. Other means were used to ventilate these barns.

When the floor of the barn was completed, the carpenter and crew commenced laying out the structural members. Rather than building the barn by installing the upright posts one at a time into the sill and tying them together with bracing, they did the heavy framing in sections on the ground. These sections, called bents, formed the skeleton of the barn. Without ever calling it such, these early carpenters were prefabricating the barn. A bent was a complete unit of framework, fully braced and extending from the sill to the point where the roof was

**Workers use pike poles to assemble the bents for a new barn, date unknown. WHi Image ID 26257**

attached. In some of the taller barns, the
bents were erected in two sections. For
some barns the two sections, when
joined together, extended upward thirty-
two or more feet.

The real skill in constructing a barn
was measuring and putting together the
bents. Workers took great care with
measuring; cutting mortises to receive
adjoining members, and fitting the
pieces together. Where the bracing members were joined, holes were drilled
through both of them with a machine

In post-and-beam construction, the barn builder fastened the timbers together with wooden pegs. This was
considered a more reliable approach than using metal
fastening devices. Waukesha County, German barn at
Old World Wisconsin, Eagle.

that consisted of a drill bit operated by a series of gears powered by two handles. The man sat
on the beam and turned both handles in front of him, drilling holes as large as one inch in
diameter through the timbers.

Once the holes were drilled, the beams and supporting pieces were joined with pegs of
oak or hickory. In the earliest barns, the pegs had to be made at the building site. Later the
pegs could be purchased. The pegs were pounded into place with a large wooden hammer,

Using pike poles and human power,
workers carefully move the post and
beams into place during a barn raising at
Alva Paddock's farm in the town of
Salem, 1891. **WHi Image ID 37272**

called a beetle, weighing up to forty pounds. This method of fastening beams together was less expensive and stronger than iron spikes or nails. Wood against wood made the unit a totality, devoid of foreign materials such as steel spikes.

It took the carpenter and his helpers three weeks to two months or more to make ready the bents prior to the barn raising. Of course, the size of the carpenter crew and the size of the barn planned affected the preparation time.

For the earlier barns, those built before 1890 or so, the farmer usually obtained his timbers from his own or a neighbor's woodlot. He probably had them sawed at a local sawmill and cured them for several months before the carpenters arrived on the scene. For many of the barns built after 1900, the farmer bought the timbers and often had them shipped by rail to the closest town. Then he, with the help of his neighbors, hauled the timbers to the building site with horse-drawn wagons. It was not unusual for the farmer to be able to purchase beams 12 inches by 12 inches, sometimes of pine but often of hemlock, up to 40 feet long. These, of course, made excellent sills, for little splicing was necessary. When all the timbers and posts had been cut and made ready, a barn raising (see page 16) was held to erect the structure.

Whether the barn was painted was up to the farmer. Some left the barn wood to weather to natural shades of gray, but most chose to add paint.

---

## ⚜ Why Are Barns Red? ⚜

**Traveling around Wisconsin** one can find blue barns, white barns, green barns, yellow barns, and barns that have never been painted and are a rather attractive silver gray. But the majority of them, of course, are red.

Why red? Many centuries ago in Europe, farmers preserved their buildings with linseed oil that they colored inexpensively with such things as animal blood from butchering, or ferrous oxide (rust). The resulting color was a rather quiet red, not the fire-engine red we sometimes see today. It became fashionable to have a red barn, which contrasted nicely with a white farmhouse.

Europeans brought their tradition of red barns to this country. Red became especially popular at the Dutch settlements in Pennsylvania, where they built with red bricks, grew red geraniums, and painted their barns red.[1]

As they did in Europe, settlers in this country mixed their own paints, often using skimmed milk, lime, red oxide, and linseed oil to make a paint that lasted many years on farm buildings. Manufactured paints were available by the mid-1800s, and they often included lead pigments such as red lead (a lead oxide). White paint was also available, but red lead paint was generally less expensive, another reason for the predominance of red barns.

**Notes**
1. Catherine Lazers Bauer, "Why Are Barns Red?" *Grit,* July–August 2007.

If I say "barn," do you think "red"? While not all barns are red, it is the most common—and very appealing—color for a barn. Marathon County, Highway 29 and Hilldale Drive east of Abbotsford.

Farmers decorated their barns in a variety of ways. Here an enormous American flag nearly covers one end of an old barn. Oneida County, off River Road near the fairgrounds.

# 8

# CHUST FOR PRETTY

## *Barn Decorations*

With its straight lines and massive size, a simple, unadorned barn is beautiful. A barn requires no fancy cupola, no special paint job, no decorations on its sides to make it more attractive. For many people, the basic simplicity of the barn gives it its beauty.

Inside the barn, the sun's rays sift through cracks in the boarded sides and play along the walls. The massive timbers that form vertical, horizontal, and diagonal lines—all fastened together without nails—offer another perspective on beauty. Their soft brown color and worn sides testify to the countless tons of hay that slipped over their edges, dragged into the barn in summer, forked out again in winter. Of all the natural beauty inside the upper parts of these old dairy barns, the greatest beauty is the massive space under the roof. The feeling of standing on the threshing floor of one of these old barns is like that of standing in a great cathedral. In both instances there is an awesome feeling of space.

Associated with the space in the big barns are the sounds, which are particularly interesting in the spring of the year, for it is then the barns are usually empty after having stored hay for the cattle during the winter months. On a windless day, the only sounds in the empty barn are the chirping of sparrows on the hayfork track or the cooing of a pigeon that has made its nest in the cupola or on a crossbeam near the roof. But let there be the least wind, even a slight breeze, and the sounds inside the barn change. Sometimes there are whistling sounds as the wind forces itself through a crack in the barn wall. Often there are creaking sounds as the beams respond to the wind, giving a little here, taking a little there. Being inside a barn on a breezy day in spring is akin to attending a symphony in a concert hall. Standing inside a great barn listening to the sounds is a kind of participation. The listener is a part of it, within it, at one with the sound and the space.

In light of all this, some would question adding adornments to any barn. True, a barn with no decoration whatsoever, not even paint, has a beauty all its own. The soft grays of aging wood on an old unpainted barn create a subtle, intensely beautiful effect. Yet many farmers chose to add color and other decorative elements.

# Paint

Most farmers, at least those who could afford it, painted their barns. The traditional pattern was red with white trim. As one travels around the state, however, many other barn colors may be observed—dark gray, white, yellow, blue, green. Probably the most popular color for barns, after red, was white. For some farmers white signified cleanliness, for others, prosperity—white paint was more expensive than red.

**Although practicality was utmost for farmers, some painted their barns white as a quiet way of saying they were doing a little better than a neighbor with a red barn. Chippewa County, County Road X east of 220th Street near Highway 29, east of Chippewa Falls.**

Some farmers painted designs on the threshing-floor doors, such as a huge white X. Some went further and painted a white circle in the center of the X. A few barns were painted in red and white vertical stripes. The Pennsylvania farmers who display hex signs on their barns today say they are "chust for pretty" and are present for no other reason. As for their historical significance, hex signs are said to have been brought to Pennsylvania from the Rhineland by Mennonites and Amish in the seventeenth century.[1] The signs were believed to ward off various problems associated with dairying, particularly cow fever. Hex signs never became popular as decorations on Wisconsin barns. Only an occasional one may be found, and it was probably purchased in a Pennsylvania craft shop.

# Lightning Rods and Weather Vanes

The lightning rod serves as protection first and decoration second. Because the big barns are so tall, they are natural targets for lightning strikes, just as is a tall tree standing alone in a field. Benjamin Franklin discovered and experimented with the first lightning rods in 1752. The idea is the same today as it was then: place a metal spire on top of a building, attach to it a metal ground cable that leads from the rod to the ground, and bury the cable several feet in the ground. That's it.

Although on many buildings lightning rods are made as simply as possible—the casual observer may not even see them—some lightning rods on barns have glass bulbs that add a splash of color to the rooftop. Beyond being decorative, the bulbs have a purpose: a broken bulb tells the farmer his barn has indeed been struck by lightning. Many such lightning rods are still found on Wisconsin barns today.

On most barns, one of the lightning rods supported a weather vane. The weather vane, sometimes known as a wind vane or a weathercock, was an important instrument to the pioneer farmer and is a useful device even today for farmers who do some of their own weather forecasting. The weather vane also was the source of folklore evidence for determining if the farmer should go fishing: "Wind in the east, fish bite the least. Wind in the west, fish bite the best. Wind in the south blows the bait right in the fish's mouth." For some reason north apparently had nothing to do with fishing. Maybe the reason it was not included is that the wind seldom blows from the north during the summer.

The traditional weather vane consisted of a turning part shaped like an arrow—one end was pointed, and the other end was wide to catch the slightest breeze. Beneath the turning part, the four directions usually were noted, although on many the direction letters were absent. Farmers knew the directions without having to look at their barns to check. Many weather vanes had figures of horses, roosters, or cows on the wide end of the arrow. In early

In addition to helping predict the weather, weather vanes provided decoration for many barns. Dane County, University of Wisconsin–Madison dairy barn, Madison.

times the rooster was a popular figure on the weather vane, thus accounting for the name weathercock. On Wisconsin barns, though, the weather vane is most likely to feature a cow or a horse.

## CUPOLAS

One often sees Wisconsin barns as well as other farm buildings proudly displaying cupolas. Cupolas are wooden structures, usually placed at the center of the barn roof. They come in various shapes. Some of them look like doghouses. Some cupolas have pointed roofs; others have simple gable roofs. Some are octagonal, some rectangular, some square. Some cupolas are very simply built; some are extremely ornate, with louvered windows, wood carvings, and elaborate designs.

According to historian Eric Sloane, the first cupolas were domed turrets used for decorative purposes. That is not the reason they were built on barns, however. The casual observer could quickly reach the conclusion that cupolas were built solely for looks, for many cupolas are quite ornate. Basically, though, no matter how simple or ornate, cupolas served first as ventilators and only second as decorations. Early ideas about ventilation—they still apply

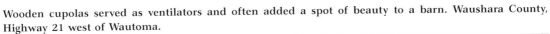

Wooden cupolas served as ventilators and often added a spot of beauty to a barn. Waushara County, Highway 21 west of Wautoma.

today—were based on the principle that warm air rises. A challenge with many early barns, particularly bank barns, that housed cattle in a partially underground basement was to provide an escape for the hot, humid air generated in the tightly enclosed area. Cupolas were built as one answer to that problem.

**Metal cupolas have their own kind of beauty, mixing line and form. Oneida County, off River Road near the fairgrounds.**

In most barns the cupola had no ducting system; the warm air simply rose into the loft of the barn and up the side walls and then moved along the roof until it escaped out the cupola. Other barns used partial ducting. Ducts leading from the stable to the loft allowed the warm air to escape more directly. Often these ventilator ducts served another purpose, as chutes for throwing down hay from the loft. The amount of warm air allowed to escape could be controlled by opening or closing the sliding door at the bottom of the chute. A more elaborate system used in some barns was a ducting system that tied the stable directly to the cupola on the roof. This was obviously a more expensive system to construct, and it could not serve a double purpose as a hay chute because it was entirely enclosed. But as a ventilation system, it was the most efficient, for the warm air could flow directly from the stable to the outside. Many of the early barns did not have cupolas, and barns being built today don't have them either. Present-day barns are ventilated with electric fans, which force the warm, humid air to the outside.

Beginning in the early 1900s, many farmers erected metal structures on their barn roofs. Though they were often about the same size as cupolas, they are more properly called ventilators. The metal structures did not have the aesthetic appeal of the wooden cupolas and were placed on the barn roof for one purpose only: ventilating the barn. Many of the metal ventilators, however, were made ornate with metal scrolling, and often a weather vane adorned the very top.

Franklin H. King, a University of Wisconsin College of Agriculture researcher, after studying Wisconsin dairy barns and noting their inadequate ventilation, developed what became known as the King system. This system, based on the knowledge that warm air rises, advocated designing barns so the warm air that rose upward from the stable could exit the structure with a ventilator of some type.

One of the popular ventilators in the early 1900s was the King Aerator. An ad for this ventilator said:

> In it [the King Aerator] are combined the principal features of construction putting into practical, successful operation the great principles of scientific ventilation. It is a galvanized steel ventilator, built of the very best materials, by the very best workmanship, and embodying in every detail the essentials of a high grade article. Note that it is highly ornamental not only in shape but in its artistic decoration. Architectural beauty and the satisfaction it gives the eye is no small item in modern building values.[2]

## WHITEWASHING

Inside the typical Wisconsin barn, little if anything was added that might be called decorative, unless whitewash on the stable walls could be called a decoration. It was common to whitewash the stable walls and ceiling as a way of increasing the available light and adding to the feeling of cleanliness.

In most of the old barns, floor joists for the barn floor above were exposed. The cracks and crevices made an excellent place for spiders to spin their webs, as they did in abundance. The spider webs naturally caught any dust in the air. And in a dairy barn, dust was more usual than unusual, particularly during the winter months, when dry hay was fed to the cattle. After a winter of spider webs and dust, the ceiling became quite dirty.

So, usually in the spring, the farmer and his children cut willow branches, tied handkerchiefs around their necks, and brushed down the dirty cobwebs some warm day after the cattle had been turned out to pasture. Once the cobwebs were brushed down and swept up from the floor, the stable was ready for whitewashing. Many farmers owned a whitewash pump, a simple device that pumped like an ordinary water pump and fit within a wooden barrel.

In the barrel the farmer mixed the whitewash, made of slaked lime and water. While the water was poured into the barrel, the mixture was constantly stirred. Heat was given off in the process. If the farmer chose to paint the whitewash mixture onto the walls, he mixed the whitewash so it was rather thick and would stick to a broom or a large brush. This was more often done with barnyard fences than with barns, however. It took a long time to paint whitewash on the exposed floor joists of the stable ceiling. Once the mixture was prepared in the barrel, one of the farmer's children or his wife worked the handle of the pump while the farmer sprayed the whitewash mixture on the walls and ceiling of the stable. Within a few minutes, the farmer was as white as the surface he was spraying.

When the whitewashing was finished, the inside of the barn was a striking white, and it smelled fresh and clean. The whitewash also discouraged spiders from returning for some time, and the barn remained sparkling for many weeks. With the advent of insecticides, certain chemicals were mixed with the whitewash, further discouraging spiders from spinning webs in the ceiling.

In some neighborhoods a custom whitewasher came by and for a few dollars sprayed the inside of the barn with whitewash. He also helped sweep down the cobwebs and helped make the stable ready for whitewashing. The custom whitewasher's sprayer was powered by a gasoline engine, eliminating the job of working the pump by hand. Though the main reason the farmer whitewashed the stable of his barn was to make it brighter and more sanitary, a newly whitewashed cow barn was undeniably pleasing to see and to smell.

In a very few instances farmers painted the insides of their barns. This was clearly an exception, usually seen only in barns owned by the so-called rich farmers or by those who, for one reason or another, wanted their barns to be showplaces.

## MURALS AND BILLBOARDS

Starting in about the 1920s and for many years thereafter, many barns, especially those located along major highways, were decorated with advertising. Bull Durham tobacco, King Midas flour, Mail Pouch tobacco, and Miller beer were among the products advertised by huge signs painted on the sides or ends of barns to capture the attention of passing motorists. Occasionally silos served as billboards as well.

Often the salesperson was able to convince the farmer to have a sign put on his barn by offering to paint the remainder of the barn or to give the farmer a case of beer or a sack of flour, depending on the product advertised. Though it is still possible to see remnants of barn advertising around the state, most firms stopped painting ads on barns during the 1950s. The

Barns have long been used as billboards, advertising everything from the county fair (as in this photo, circa 1897) to chewing tobacco, beer, and the circus. **WHi Image ID 56104**

Peavy Company, the Minneapolis manufacturer of King Midas flour, hasn't commissioned any barnside ads since the early 1950s. At that time the company actually paid to have some of the ads painted over.

Barnside advertising was always controversial. Many people disliked seeing an otherwise-beautiful barn adorned with commercial messages. The Interstate Highway System and highway beautification programs contributed to the practice's demise.

A somewhat more respectable type of barnside decoration is a mural. Frank Engebretson was the best known of the Wisconsin artists who painted barn murals. His work was recognized nationally when a photo of one of his murals appeared in the August 28, 1942, issue of *Life* magazine.

In the 1970s barnside mural painting was revived through the efforts of the Wisconsin Arts Board and its Dairyland Graphics project. They selected several barns around the state—in most cases ones that were easily visible from a highway—and local artists teamed with local youths to paint murals on them. The Arts Board financed the project through a grant from the National Endowment for the Arts and donated paint and supplies. The project began in the spring of 1975. By October 1976 twelve barns scattered throughout the state had murals on their sides, their ends, or both.

Members of 4-H, Future Farmers of America, and other youth groups did most of the actual painting under the careful guidance of a local artist. The local artist, in turn, received direction from a muralist employed by the Wisconsin Arts Board. The muralist had experience with the Chicago Mural Renaissance, which sets the national standard for all community mural programs.

Not everyone was thrilled with the Dairyland Graphics project. Just as many people disliked advertising on the sides of barns, many disapproved of murals on barns. "Why can't we be satisfied with the natural beauty of barns? Why do we feel a compulsion to paint pictures on them?" asked a sizable number of people. These critics had a point. Today few of these murals remain. Most have faded with time or have been painted over.

## ⚹ Barn Quilt Designs ⚹

**Folks in Ohio launched the first barn quilt project in 2001,** and the idea has spread throughout the Midwest. Green County was the first to start the project in Wisconsin; as of this writing the group has more than sixty barn quilts completed or in the works. The quilt artists develop 8-by-8-foot designs that are then painted on wood blocks by volunteers and mounted on the sides or ends of barns. According to project cochair Lynn Lokken, "The barns of Wisconsin are important to agriculture, and quilting was and still is a creative and social outlet for women. I think they are a perfect match."[1] Similar barn quilt projects are currently under way in Walworth, Monroe, Door, Kewaunee, Lafayette, and Racine Counties.

**Notes**
1. Personal correspondence, Lynn Lokken, July 13, 2008.

Quilt art, like this Swiss star pattern, is now found on barns in several Wisconsin counties. Green County, N4512 Cold Springs Road, Monroe.

An example of a barn mural with a humorous twist. Dane County, Highway 19 near Sun Prairie.

Built in 1897, the University of Wisconsin–Madison's dairy barn was listed as a National Historic Landmark by the U.S. Department of Interior in 2005.

# 9

# SOMETHING SPECIAL

## *Unusual Barns*

Deciding what is a "special" barn is a difficult task, for the word *special* holds different meanings for different people. One person's special barn may be quite ordinary to another.

Barns are quite special to most people who till the soil. The buildings obviously are of economic importance as shelters for animals and feed, but they also hold deeper meanings for farmers, whose parents or grandparents probably had a part in building the structures. For many farmers, their barns have become symbols of their heritage and expressions of their rural life. In this sense, there are thousands of special barns scattered throughout the state. Those discussed here were chosen for their fascinating stories, which, luckily for us, are part of the historic record.

These unique barns fall into several categories. One is the barn that was built with little concern for cost and with an emphasis on the ideal. The barn at the Green Lake Conference Center is an example. Another category is a barn built for a special purpose. The old dairy barn on the University of Wisconsin–Madison campus fits here. Still another category of barn is one that was once used as a barn but was later converted to another use. Many examples of barn conversions exist throughout the state. Eating establishments seem to be popular uses for old barns; other old barns have been remodeled into homes.

Then there are the barns originally built for another purpose. A pioneer's log home was often converted into a barn. Some of these log barns, once cabins, were used for years. Even after the farmer built a frame barn, the log structure often continued to house young stock—pigs, sheep, or sometimes chickens. Back in the early 1900s, before the coming of the automobile, rural churches stood at many crossroads. When these closed, some of them were moved to neighboring farms where they became barns. The same fate befell some rural one-room schoolhouses.

## Lawson's Guernsey Barn

Victor Lawson, former publisher of the *Chicago Daily News* and a founder of the Associated Press, was influenced by his wife to buy land on Green Lake in central Wisconsin. As the story goes, one day in the summer of 1888, Jessie Lawson took friends for a boat ride on Green Lake. When a sudden storm came up, they were forced to put in on Lone Tree Point, on the north side of the lake. They took shelter in a shack they found there.

Upon returning to Chicago, Jessie Lawson told her husband they must buy the land where she had taken refuge from the storm. Before the end of the year, Lawson had purchased nearly ten acres of land, including Lone Tree Point, for five hundred dollars. This was only the beginning. The Lawsons added several farms to their estate until they had accumulated more than eleven hundred acres.

Jessie Lawson took primary interest in developing the estate. The Lawsons spent several million dollars building a spacious home, twelve miles of paved roads, two sets of farm buildings, a boathouse, two greenhouses, a powerhouse, seven water towers, a small golf course, and homes for the workers. The Lawsons maintained horses, pigs, sheep, and fine herds of Guernsey and Jersey cattle.

**The former Lawson dairy barn is now known as William Carey Hall and is part of the Green Lake Conference Center in Green Lake County.**

The estate continued in the hands of the Lawsons until Victor's death in 1925; Jessie had died in 1914. In 1925 the property was sold to the H. O. Stone Company of Chicago. This company spent another three million dollars on what came to be known as Lawsonia. The stock market crash of 1929 and the Depression succeeded in putting Lawsonia into receivership. The bank holding the mortgage operated Lawsonia for about ten years, but it, too, was forced to close the gates.

In 1943 the then Northern Baptist Convention, now the American Baptist Churches/USA, was looking for a place to use as a national convention center. They heard about this "fabulous fool's paradise" in Wisconsin. By the end of the year they had purchased Lawsonia for $300,000 and realized their dream for a truly outstanding site for a conference center that would focus on spiritual growth, leader development, and recreation. The farm buildings, including the Lawsons' former Guernsey barn, were converted into conference facilities. Today the barn is known as William Carey Hall, named in honor of an eighteenth-century English Baptist missionary to India.

The Lawsons' Guernsey barn was constructed by W. A. Merigold Jr. in 1915. The main section of the barn is 40 feet wide, 225 feet long, and 45 feet high from the top of the basement floor to the top of the ridge. The barn is of wood-frame construction, with a concrete foundation. The frame consists of a series of trusses spaced 14 feet apart. The area between trusses is framed with 2-inch by 6-inch girts, spiked and bolted together. The truss timbers are of pine and hemlock, 2 inches by 10 inches, and 2 inches by 12 inches.

The exterior of the barn is covered with two layers of wood boards. The first layer is matched 8-inch sheathing applied vertically. The second layer, exposed to the weather, is a drop-siding pattern board applied horizontally. Building paper was laid between the two layers of siding. The gambrel roof was covered with roof boards and cedar shingles.

The floors and gutters of the stable section were constructed of poured concrete, except for the stalls, which received special treatment. The Lawsons were concerned about the feet of their prize dairy animals, so they installed cork brick on the stall and pen floors to provide more comfortable quarters for the cattle. All the gutters and pens were equipped with drains that ran the entire length of the barn, to an outside holding tank. The contents of the tank were pumped out and used for fertilizer. A steel track ran the length of the barn, behind the stalls, permitting the use of a manure carrier for the removal of solid wastes.

The floor plan for the lower level of the barn included two rows of cow stalls, each row having a capacity of fifty cows. The rows of cows faced each other to simplify feeding. The feeding cart could be pushed down the alley between the two rows, and feed could be distributed on either side. Each cow stall had an automatic watering cup. In addition to the two rows of cow stalls, there were two calf pens, a pen for calving and illness, and a bull pen, all located toward the rear of the barn.

The extreme south end of the barn was partitioned to provide a cream separator and a combination locker and feed room, as well as access to the silos. Two clay-tile silos stand at the south end of the barn, each 24 feet by 46 feet, with 4-foot pits and a capacity of about 250 tons of corn silage.

The hayloft of the barn measured 34 feet 6 inches from the mow floor to the rafters. The loft was equipped with a track and hayfork and had storage space for about 410 tons of loose hay, in addition to granary space for feed grains. The hayfork doors, which opened to the outside of the barn, were unique because of their large size, their design, and their many panels of glass.

In addition to the main section of the barn, there are two large wings running east and west from the main structure. Each measures 36 feet wide, 137 feet long, and 24 feet high and was equipped with track and hayfork on the upper level. The lower level of the two wings was designed for young stock and heifers. Much of the flooring used in the wings was cork brick, the same type used in the main barn.

Several features of this barn made it unusual. Concern for natural light required the builders to install many windows on the east and west sides; eighteen small dormers in the upper part of the barn provided light and additional ventilation to the haymow. The barn was also well ventilated by a gravity-flow system, based on the principle that cold air settles and warm air rises. Four large metal stacks vented the warm, moist air out through the roof.

But probably the most unusual feature of the barn was its interior finish. The ceiling and all the interior walls were covered with four-inch V siding of yellow pine, the same material used in many houses of the day. The windows and door openings were cased just as in a house, and the entire interior was varnished, as were some of the finer houses built during that period. The quality of construction was truly outstanding throughout the entire barn, making it a showplace among dairy barns in the state.

According to the *Guernsey Book of Registry*, among the historical material owned by the American Baptist Assembly, the Guernsey show herd was sold in 1923. During World War II the barn housed German prisoners of war, many of whom worked in the canning factories in the area. Once the Baptists acquired Lawsonia, they began remodeling many of the buildings on the estate, including the farm buildings. The loft of the Guernsey barn was used for many years as an auditorium. It seated eight hundred people, had excellent acoustics, and included a stage, dressing rooms and wardrobe rooms, and an electronic organ. When the barn loft was converted into an auditorium, the silos on the south end of the barn were converted into restrooms. The loft is currently used primarily for storage.

The Baptists remodeled the west wing of the barn into housing for youth conferences and converted the east wing to a kitchen and dining room. Currently the west wing has been remodeled into apartments, housing the adult volunteer staff. The east wing was demolished in the late 1990s, and a new maintenance building for the Lawsonia Golf Course was constructed on the site. The new construction is in an architectural style consistent with the original barn.[1]

In many ways this early-twentieth-century dairy barn was the same as many other dairy barns being constructed in Wisconsin at the time. The gambrel-roof design was frequently used then, metal ventilators were often erected on the roof, and wood construction was common throughout. Where this barn differed from the ordinary barns is its size, the elegant way it was constructed, and the extras that were included, from the cork-block stall floors to the house-type framing for windows and doors.

## THOMAS STONE BARN

Located in Iowa County, between the communities of Ridgeway and Barneveld on State Highway 18–151, the Thomas barn is built of quarried limestone and is 100 feet long, 40.5 feet wide, with limestone walls from ground to roof. The rock for the barn was quarried on the farm and thus needed to be hauled but a short distance using stone boats. Another unique feature of this barn is the stable's bedrock floor.

**Built in 1881 and owned by the Thomas family for four generations, this barn is constructed of stone taken from the farm's quarry. The barn is on the National Register of Historic Places. Iowa County, Highway 151 between Barneveld and Mount Horeb.**

Welsh immigrant Walter Thomas designed the barn, and three Welsh stonemasons built it in 1881. The quarried rock wall is generally 21 inches thick. The arched doors and windows are set with keystones and little mortar.

The two-story barn has the stable below and a large hay storage area above. The barn, constructed similarly to bank barns, allows for ground-level entry to the stable area. The stable area includes ten doors that open to the south, an unusual feature. The north facade of the barn, which faces the highway, has two large, arch-topped doors, and another door is located on the barn's west end. The building's unique construction features—quarry-stone walls; round-arched, drive-through doorways—have not been altered since it was built. The barn is one of a few in the state that were constructed totally of quarried rock, and the only one of its kind in Iowa County.

Walter Thomas was involved with beef farming, having at one time eight hundred head of cattle. His new barn provided a much-needed storage area for winter feed. Thomas converted the lower level of the barn to house dairy cows in 1912. The farmstead borders 85 acres that the Harold Thomas family sold to the Driftless Area Land Conservancy in 2005. Some of these funds from the land sale helped to restore this magnificent old barn.

The Thomas barn has remained in the Thomas family for four generations. The building was added to the State Register and National Register of Historic Places in 2001.[2]

## UNIVERSITY OF WISCONSIN'S COLLEGE OF AGRICULTURE BARN

The University of Wisconsin College of Agriculture became a leader in promoting dairying in Wisconsin in the late 1800s. With increasing demands for teaching and research about dairying, a new dairy barn for the college became a necessity. In 1896 William A. Henry, dean of the college, visited a number of agricultural colleges and large farms in the eastern states and in Canada to gather ideas about building such a structure.[3]

In 1897, after sufficient funds had been appropriated for a new barn, Dean Henry hired architect J. T. W. Jennings of Chicago to draw the plans. Jennings had designed the UW's King Hall and the Agricultural Heating Station. For the UW's new barn, Jennings chose an architectural style that was seen in certain parts of Normandy, France, at that time and had attracted the attention of tourists. Because the barn was to become an agricultural college building, several modifications were made that would not have been included in an ordinary farmer's barn. In addition to housing cattle for teaching and research purposes, the barn needed space to accommodate large numbers of students in a classroom setting. The interior of the barn was planned by agricultural faculty and staff in consultation with construction experts.

The University of Wisconsin–Madison's dairy barn (seen here circa 1904) included an elevated ramp (at right) that allowed teamsters with wagons to haul loads of hay to the upper haylofts. **WHi Image ID 55522**

Dean Henry, in his fifteenth annual report (June 1898) of the Agricultural Experiment Station, wrote:

> During the past year a barn for dairy cattle furnished with commodious quarters for judging live stock by our students has been completed on the University farm at a cost of about $16,000, with $2,000 additional for equipment. Our Agricultural College now has a dairy barn which is worthy in some measure of the great dairy industry pursued by our people, and in the room devoted to stock judging we have the comfortable quarters so much needed by the students of the Agricultural College.[4]

The building is a wood-frame structure consisting of the three-story barn proper, 86 feet long by 50 feet wide, with two two-story wings, one 70 feet long by 40 feet wide and one 70 feet by 30 feet, projecting at right angles from each end of the south side of the main building. The large classroom area, 70 feet by 40 feet, was located between the two wings and was lighted from the roof as well as from the gable end. The classroom included seats along the west side and a tanbark floor. A boiler in the basement provided heat for the room through coils of steam pipe.

A six-span steel trestle, no longer present, was built so horse-drawn wagons could be driven directly onto the third floor of the barn. The grade of the trestle was moderate, 7 feet in 100 feet, so a team of horses could easily pull a load of hay, grain, or corn fodder into the barn. All cattle feed was hauled to the third floor, but, with the exception of hay, no feed was stored there. Thus the third floor was mostly an open area, large enough so a team and wagon could enter, unload, and turn around.

A wagon scale was located on the west end of the third floor so all feeds coming into the barn could be weighed before being stored. A feed grinder stood on the southwest corner; a silage cutter was on the northeast corner. Both of these machines were powered by the same ten-horsepower electric motor, which was fastened to a small carrier that ran on an iron track so the motor could be shifted easily from one side of the barn to the other. This eliminated the need for a complicated system of belts and pulleys.

When the feed grains were ground, they were dropped through trapdoors into bins on the second floor, which was a storage area, divided into hay-storage bays, space for corn fodder, and several grain bins. Five grain bins were designated, one each for oats, oil meal, bran, shorts, and cornmeal.

The first floor of the main part of the barn had a large driveway with doors opening to the north. Three hospital stalls and a hay-storage area were to the west of the driveway. The herdsman's office and a small bedroom were to the east of the driveway. The feed room, with ducts leading from the grain-storage areas on the second floor, was across an alleyway. A milk room that had a tile floor and a marble baseboard was attached to this east end of the barn. The inside of the milk room was lined with steel sheeting heavily coated with white porcelain paint. A small cream separator powered by a one-horsepower electric motor was included in the room. The manufacturer had nickel plated both the cream separator and the motor at no cost to the college.

The east wing of the barn, 70 feet long by 40 feet wide, included stalls for thirty-six dairy cattle and was lined with corrugated galvanized-iron sheeting, which in those days was considered the most sanitary building material because the entire surface could be washed with hose and brush when necessary. The center passageway in the cow stable was ten feet wide so a team could be driven through the barn to feed green crops directly from the field.

The west wing, 70 feet long by 30 feet wide, housed young stock and bulls and was divided into several pens. It, too, was lined with galvanized iron. The floors of both wings as well as the mangers were made of Portland cement and crushed granite. They sloped slightly so all water used in washing ran to the sewer drains.

The basement, located under the main part of the barn, was divided into two major rooms. One was used for the storage of roots and other produce. In the other was the heating plant, the farm workshop, and an artesian well. A three-horsepower electric motor pumped water into a large steel tank located above the silo. A system of underground pipes distributed water to all the farm buildings from this water tank.

Agricultural physicist Franklin Hiram King designed a silo that was constructed near the northeast corner of the barn. It was filled by dumping cut corn fodder into the top of the structure. The silo is 18 feet across, inside diameter, and 33 feet deep. It is of frame construction, lined inside and outside with brick. The inner brick surface was heavily coated with Portland cement, making it essentially watertight.

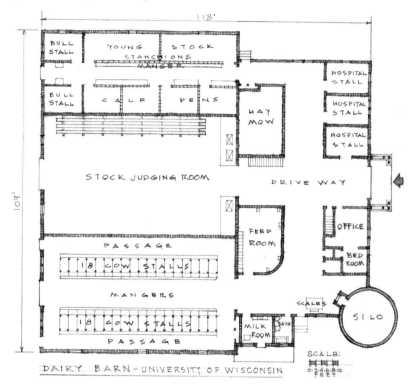

**Floor plan of the University of Wisconsin dairy barn.** Drawing by Allen Strang, courtesy of William Strang

F. H. King's ventilation system consisted of a flue, a type of cupola, built on the roof of each wing of the barn. From the cupola, two ventilating shafts ran along the roof of each barn and down the sides to the stable. Openings at the floor and near the ceilings of the stable could be regulated to allow more or less warm air to escape into the ventilation system. Provision was made so cold air could enter the stable at various points in the barn's side walls. All reports suggest it worked well.

The dairy barn was a showplace on the college campus for many years. Not only had it been designed as a functional building for the combined purposes of housing cattle and providing for classroom teaching and college research, it was an attractive building. No farmer would ever build a barn like this one, yet it provided farmers with many ideas about barn ventilation, construction of cow stalls, and feed handling.

Later the university added two more wings to the barn, one on the east and one on the west. In 1954 a one-story dairy barn of steel-and-brick construction was built just to the east of the original barn. Though the iron trestle that led to the third floor has been removed, the old barn and the original silo stand essentially as they were constructed.

As the years passed, less and less attention was paid to the barn. In 1946, with a great influx of World War II veterans on campus, a committee inventoried campus buildings and made recommendations about their maintenance and retention. The committee, chaired by A. F. Gallistel, superintendent of UW Department of Buildings and Grounds, said that the dairy barn was obsolete and an extreme fire hazard. The committee recommended the barn be razed when a replacement became available.[5] But somehow the old dairy barn survived. With the renewed emphasis on historic preservation, especially following the U.S. bicentennial in the mid-1970s, interest in keeping the barn developed.

The old barn was the site for groundbreaking research on animal nutrition that eventually led to the isolation of vitamin A in 1913. These experiments, tracing back to the early 1900s, led to profoundly different approaches to feeding cattle compared to what had been previously known. Other research conducted included breeding advances through artificial insemination, teaching and testing techniques for bovine tuberculosis, an improved test for brucellosis, and discovery of causes for milk fever.

In 2002, with the sponsorship of the Barns Network of Wisconsin and the Madison Trust for Historic Preservation, the UW dairy barn was placed on the National Register of Historic Places. And on April 5, 2005, the secretary of interior listed the UW dairy barn as a National Historic Landmark. This is the highest award given to a qualifying structure of place in America. It is the only barn on the National Historic Landmark list.[6]

The old dairy barn, now used mainly for storage, stands today as a reminder of the University of Wisconsin's role in providing leadership to the emerging dairy industry in Wisconsin.

## VOEGELI BARN

The Voegeli farm is located just south of New Glarus, in the hills of Green County. Traveling along Highway 69, one cannot help but notice the huge white barn that contrasts with the more typical bank barns with pentroofs that are often found in this part of the state.

The Voegeli family has farmed this same land since 1853, when the first Voegeli arrived from Switzerland. Until 1917 the Voegelis had a bank barn with a pentroof. Built in about 1860, the old barn was still quite sturdy when they tore it down to make way for the present structure. "It was quite a decision my father and I had to make," said Jake Voegeli, who was eighty-three when he was interviewed in 1976. He described how the family decided to replace the old barn because it was too small for their rapidly expanding herd of prize-winning Brown Swiss cattle.

The same carpenter crew of eight to ten men that built the new barn also tore down the old one. The two jobs took most of the summer of 1917. Because there was no power-driven equipment in those days, they did everything by hand.

From the beginning, the nature of the building site was a problem. The old barn had faced south, and the new one would be in the same place. But the site was on a rather steep hillside, with a creek at the bottom of the hill. The first task was to build a retaining wall of limestone, which is native in the area. The same type of rock was used to construct the three-foot-thick barn walls, which in places are 27 feet high.

Many of the materials from the old barn were used in the construction of the new one, from the huge, 10-inch by 10-inch posts used in the new haymows to the old barn siding used as roof boards in the new barn. Timbers of the size required for the new barn were not available in Wisconsin, so they were shipped in from Washington State strapped to two railcars. The timbers, which were used as beams to support the main part of the barn, are 12 inches by 12 inches by 52 feet long and cut from fir trees. The tie beams are also 52 feet long, some 6 inches by 12 inches, some 8 inches by 8 inches.

Because of the unusual building site—none of the back portion of the barn's stable is above ground—the ceiling of the stable is 11 feet high, so sufficient light can enter from the south, where the windows are located. Even on a cloudy day the barn is well lighted without the use of electricity. Also, the high ceiling makes the task of ventilation somewhat easier.

Originally the barn was built with ventilators on the roof. Round metal shafts connected the roof ventilators with the stable. But, according to Voegeli, this system never worked well. Later the barn was ventilated with exhaust fans in the side walls.

The overall dimensions of the barn are 52 feet wide by 144 feet long by 59 feet high from the stable floor to the peak of the roof. The width of the barn allows for three rows of live-stock—two rows of cattle arranged with tails facing and an additional row consisting of calf pens and eight tie stalls. The arrangement includes two walkways.

The haymows above the stable are 16 feet wide. One mow has a driveway that is entered from ground level on the north side of the barn, but because the hillside is so steep, the

Built in 1917 with 12-inch by 12-inch timbers 52 feet long, the Voegeli barn has been retrofitted for a milking parlor. Green County, N7190 Highway 69, Monticello.

## ⚜ Circus World Museum Baggage Horse Barn ⚜

**From 1884 to 1918 the Ringling Brothers Circus spent the winters in Baraboo.** The troupe's winter quarters, on the banks of the Baraboo River, consisted of a collection of buildings that provided shelter for the animals that were part of the show, from tigers and lions and huge snakes to camels, elephants, and horses.

Percheron draft horses did the hard physical labor of the circus before trucks and tractors became available. These baggage horses pulled the colorful circus wagons in parades, assisted in setting up the tents, and provided most of the power necessary for moving and running a circus. As the Ringling brothers grew their circus, they steadily increased their number of horses.

Wisconsin's harsh winters meant horses needed ample shelter, and by 1903, with more than three hundred baggage horses to feed and house, the Ringlings had run out of stables. In 1904 they hired contractor Carl Isenberg to build a new horse barn on Water Street in Baraboo. The barn was 56 feet wide and 120 feet long, and it housed about one hundred horses. It cost $4,500 to build.[1]

The barn included individual mangers and troughs for feed and water, and each horse's stall floor had wooden planking, which was far less damaging to the horses' hooves than concrete. One corner of the barn housed a small work area where a blacksmith shod horses.

Today the baggage horse barn, along with several other original Ringling winter quarter buildings, are a part of Circus World Museum, operated by the Wisconsin Historical Society and open to the public. A visit to this unique barn can bring back the sights and sounds of the circus, when horses did much of the heavy work around the circus grounds as well as performing in the ring.

### Notes

1. Jerry Apps, *Ringlingville USA: The Stupendous Story of Seven Siblings and Their Stunning Circus Success* (Madison: Wisconsin Historical Society Press, 2005), 52–53, 115.

The Ringling brothers' baggage horse barn, built in Baraboo in 1904, housed about one hundred Percheron horses.

driveway is 11 feet above the level of the mow floor. Under the driveway are bins for storing grain and sawdust, all fed by gravity to the stable below. The hip posts in the mows are 27 feet long. The entire framing for the barn was built without nails. Beams were mortised together and then made secure with wooden pegs. The beams were carefully spliced with an intricate notching system that was shaped by hand so the pieces fit perfectly.

The Voegeli barn was remodeled in the summer of 2008. A milking parlor was retrofitted into the existing cattle housing area. The tie stalls and stanchions used to hold the cows during milking were removed and replaced with a holding area. The space also allows for an office and special needs pens.[7]

Glazed tile silo with a quarried rock foundation. Some believed silage was less likely to freeze in this type of silo, as the tiles had air spaces within them that provided some insulation. La Crosse County, County Roads C and T.

# 10
# TOWERS OF FERMENT

## *Silos*

In some ways, the silos of Wisconsin define the state and its dairy history as well as do its dairy barns. All around the state these grand silos serve as exclamation points on the rural landscape.

As a fourteen-year-old farm boy it was my task to steer the end of the silo filler pipe into the little window that stuck out from the roof of our silo. The task terrified me because I disliked high places, yet from this vantage point I had an amazing view. I could look off across the fields and see the neighbor's farmstead in a cluster of trees to the south. I could see our cornfields and hayfields, and the stubble that had been a field of oats. I could see Holstein cattle grazing in the pasture in the far corner of our farm, more than a mile from the barn. In many ways, I could see the world from my precarious perch at the top of the silo.

As I clung to the top rung of the ladder, I could hear the sparrows that nested under the roof. Their chirping song was amplified by the huge black expanse of the empty silo. With one hand I wired the filler pipe into place and climbed slowly down the ladder.

Neighbors gathered on the day my father had set aside for silo filling. They hauled loads of green cornstalks to the silo filler, which was pushed against the silo wall. I could smell the rich aroma of harnesses, sweaty horses, and fresh-cut corn as the tractor powering the filler sputtered to life.

"Whoa!" the driver yelled to his team when his wagon was next to the filler. The horses shied a bit when they pulled alongside the noisy machine. The teamster, his shirt soaked with sweat, climbed off the wagon and began shoving stalks into the ever-hungry silo

**Farm workers feed corn into a tractor-operated silo filler, circa 1916. WHi Image ID 46769**

filler. The machine chopped the corn and blew the pieces up the long filler pipe into the silo, where they rained down like hail.

The neighbors fed load after load of stalks into the machine until, late in the afternoon, I pushed a handful of cut corn out the silo window, signaling the silo was full.

The crew filed into the house, where the table was spread with heaping bowls of steaming mashed potatoes, platters of canned meat, stacks of homemade bread, and at least three kinds of pie plus a cake or two. The next day the neighbors returned for a couple of hours to refill the silo because overnight the corn had settled several feet.

For many years this was the procedure used to fill silos on Wisconsin farms. Today, a forage harvester, often self-propelled, cuts the corn into the proper silage length right in the field. The loads of cut corn are hauled to the silo, where a machine blows the corn up the length of the silo to drop down inside. The farmer never touches the corn. Gone also are the silo-filling crews, made up of neighbors who went from farm to farm until all of the silos in the neighborhood had been filled.

But the silos remain. For every Wisconsin dairy barn there is likely to be at least one silo standing nearby; some farms have as many as six or more. Over the years, silos have continued to be popular feed-storage structures, particularly for corn and, in recent years, for legumes and grasses as well. Silos come in various shapes, sizes, and colors. Some are mostly underground and consist of trenches dug into the earth. The earth sides of the trench become the walls for the silo. Most silos seen today, though, are upright cylindrical structures. Some are as small as 8 feet in diameter and 20 feet tall. Others may be 20 or more feet in diameter and 80 feet tall.

Older silos were often made of wooden staves. Redwood or cedar was preferred because these woods rotted slowly and could better withstand the action of the various acids resulting from the fermentation process that occurs in silos. Some silos were constructed of concrete—either poured concrete, concrete blocks, or concrete staves (rectangular preformed pieces of concrete that were fitted together to make the silo walls). A few silos were made of fieldstones and had walls three feet thick. Some were built of tile or brick; a fair number were constructed of glass-lined steel. And the modern-day silo doesn't look like a silo at all but appears to be a giant white grub worm about four feet in diameter and several feet long, lying horizontal on the ground. Called silage bags, these "silos" are made of heavyweight plastic. Silage is blown in the end of the bag, which is usually sealed by piling soil on the end. Farmers find them convenient and inexpensive.

Many large-scale dairy farmers have returned to the use of trench or bunker silos, which generally have concrete sides and floors and are open at the top and at one end. Silage is blown

Many of the early cylindrical silos were of wood-stave construction. The wood was usually redwood or cedar, resistant to rot. Waushara County, Highway 21 west of Wautoma.

into the structure and is packed by driving a tractor on it. Then it is covered with heavyweight plastic. Bobcats and other front-end-loader-type tractors are used to feed from these huge bunker silos, which somewhat resemble the first silos used.

Unlike expertise in barn building, skills in silo building didn't come to Wisconsin with the pioneers. This was mainly because the use of silos became popular after many of the pioneers had arrived. In fact, no silos existed in Wisconsin or in the United States until the late 1800s. An article in the January 28, 1922, issue of *Prairie Farmer* claimed that Fred L. Hatch of McHenry County, Illinois, built the first silo in North America, in 1873. Hatch's silo was not a

Common on farms today, plastic bag silos here contrast with the more conventional concrete-stave silos in the background. Iowa County, Jacobson Road and County Road Y.

cylindrical upright structure as is true for most silos today. It was rectangular, 10 feet wide, 16 feet long, and 24 feet deep, with 8 of those feet underground.

An earlier claim for the first silo in this country was made by a writer in *The Country Gentleman* of January 20, 1881. He claimed a German at Troy, New York, had built a silo in 1865. It was of stone construction, 25 feet by 30 feet by 8 feet tall. It was filled once and then abandoned because the owner was disappointed with the results.

Who really built the first silo in the United States is not clear, but in any case, it didn't occur until the second half of the nineteenth century. Levi P. Gilbert of Fort Atkinson is said to have built the first silo in Wisconsin, in 1877. His silo was a trench silo, 6 feet deep, 6 feet wide, and 30 feet long.[1]

## Early Silo History

Though silos didn't come into any kind of prominence in this country until around 1900, the practice of burying grain in underground pits, either to store it for future use or to hide and protect it from rodents or enemies, is mentioned by some of the early Greek writers. The word *silo* comes from the Latin *sirus,* "a pit to keep corn in."[2]

The man who deserves primary credit for modernizing silage making is August Goffart of France. As early as 1852 he began to study the preservation of forage, and in 1877 he published his famous book, *Manuel de la Culture et de I'Ensilage des Mais et Autres Fourrages Verts,* which summarized twenty-five years of practical experience and research. It was translated into English and published in New York during the winter of 1878–1879.[3]

The idea of preserving a crop in an underground trench caught on slowly in this country. A United States Department of Agriculture report issued in July 1882 listed ninety-one persons who had constructed silos in the entire country. Of that number, eighty-one were in eastern states. Three Wisconsin names were mentioned—Henry Lapham and L. W. Weeks, both of Oconomowoc, and the University of Wisconsin.

J. M. Bailey of Billerica, Massachusetts, built a silo in 1879 and became an advocate of this system for preserving corn. He did considerable writing about the benefits of silos and silage, including a book titled *The Book of Ensilage, or the New Dispensation for Farmers.* Bailey wrote a letter to W. D. Hoard of the *Jefferson County Union,* which was published June 4, 1880. In the letter he said:

. . . Since the opening of my silos, December 3, 1879, I have been feeding a large stock of cattle and sheep upon corn fodder, ensilaged last September. I am now feeding my milk cows and ewes with lambs upon it exclusively. Its preservation is as perfect as when the silos were first opened. Capacity of silos, 800,000 pounds.

You, the progressive farmers, dairymen, and all you readers who are interested in the preservation of green forage crops for winter feed by the system of ensilage, are cordially invited to visit "winning farm" and witness the results of the first thorough trial of this system in America to inspect the silos, the ensilage the stock fed upon, and test the quantity of the milk produced by ensilage.[4]

Bailey and others took up the silo with missionary zeal. They set out to convert their fellow farmers to this new way of preserving feed for their cattle, and they did it with such enthusiasm that occasionally they stretched the truth a bit. One convert wrote, "Since I have fed ensilage, I have not had a case of garget or other ill condition of the udder, nor a case of ill health in all the time. I look upon ensilage as a complete substitute for roots in maintaining the health of animals, and if for no other reason I would always feed half feed of ensilage in dairying for profit."[5]

Another convert, L. B. Marsh of Trempealeau County, built his first silo in 1881. He wrote, "The silo is a treasurer to the poor farmer, for with spade in hand, a silo he can have and the smaller the farm and poorer the farmer the more a silo he must have."[6] A German agricultural writer, completely carried away with his enthusiasm for the future of silos, said the day would likely come when dry hay would be obtainable only in drugstores. Livestock would be fed only silage.

Dean Henry of the University of Wisconsin's College of Agriculture asked state farmers who had silos to write to him and carefully document both the advantages and disadvantages of silage. L. W. Weeks of Oconomowoc wrote to Professor Henry on December 16, 1882.

It is a very simple story. Before 1880, I was plodding along with a few cows, my little farm of 48 acres was not a grass farm, and as to that matter, not much of any kind of a farm, being looked upon by my neighbors as no farm at all. I found it impossible to raise hay or grain enough to carry half a dozen cows. I had commenced making fine butter, adopting the Danish system of cold setting milk, of which I had learned something in my wanderings in Europe, meeting with fair success. I increased the number of cows to about a dozen, purchasing hay and grain for winter feed with part soiling. I found, however, that little or no profit remained, and it became a question whether or not to abandon the undertaking as impracticable.

Before deciding to give it up entirely, I came across Goffart's treatise on ensilage. This seemed to open a possible door of escape from the dilemma; after thoroughly investigating from every source of information within my reach, I built my silo and put in about 100 tons of fodder corn. It was not cut fine enough or stamped sufficiently solid to produce the best ensilage, but it was eaten readily, and with a liberal ration of corn meal, oil meal and mill stuffs I was able to increase the number of cows to nineteen. They did well during the winter coming out in high flesh in the spring. I bought no hay for them and had only a few tons of marsh hay grown on the farm.

During the summer of 1881 I added to my little farm by purchase twenty-five acres of excellent corn land, but got small benefit that season. I, however, put up about 150 tons of corn and rye ensilage, which allowed me to increase my stock to twenty-six head. . . . The little farm has given a liberal profit each year since I commenced to use ensilage. Before that, it paid a yearly loss. . . .

Concerning the disadvantages of silage, Weeks wrote:

To me it has had but one dark spot, and that is the great trouble of procuring and having about you a dozen men at a season when harvesting is claiming every available man, or if you plant your corn later you come into the threshing period, when high wages in either case will make the cost too much even if you can get them at all. I have been forced to employ women and children even to cut my ensilage in time. This may be of less consequence in other sections, but here it is serious, and if a dozen of my neighbors were doing the same thing, where would the labor come from?[7]

Apparently the idea of exchanging help among farmers for various farm operations was not practiced at the time Weeks wrote his letter. From the tone of Weeks's letter he was not seen as a "regular" farmer in his neighborhood. This may have accounted for some of the "dark spot" he spoke of. Nevertheless, Weeks is credited with building the first aboveground silo in Wisconsin, in 1880. Below ground level, his silo was built of stone and cement to a depth of 12 feet. Above ground, it had a wooden superstructure double boarded on the inside. Top to bottom, the silo measured about 20 feet.[8]

John Steele of Alderly, in Dodge County, got his idea for building a silo from government reports about the subject. He built a silo in 1880 and filled it with twenty-five tons of corn fodder. The Steele silo, actually a root cellar, was filled in one day, the crew working from early morning until eleven o'clock at night before they finished. Steele became a Wisconsin silo missionary. He readily gave instructions to farmers visiting his farm, and he wrote articles that appeared in the *Breeder's Gazette* and *The Country Gentleman*.

A year after he first filled his silo, Steele extended the walls of the old root cellar (which was inside his barn) up into the haymow, bringing the top of it even with the eaves. This made the silo 15 feet by 16 feet by 23 feet deep. Twelve feet were above ground. The silo was of stone construction, double boarded, with building paper used as insulation.

On November 8, 1880, Steele was elected to the state assembly. He became a member of the agricultural committee and was influential in obtaining for the UW College of Agriculture's Dean Henry an appropriation of four thousand dollars to experiment on silage and sugarcane. He was the only one in the assembly who had a silo or knew anything about silage.[9]

The American farmer is not one to try a newfangled notion unless sound evidence is available that it will work on his farm. Even with all the excitement generated about the virtues of silos, they did not spread quickly across the land. In fact, a considerable amount of talk started that challenged the claims of the silo disciples. Some farmers argued that feeding silage would cause cows to lose their teeth, that it would eat out their stomachs, and that it would cause trouble at calving time. Most of these farmers, however, did not own silos. One farmer who did have a silo claimed silage made his cows drunk, causing them to stagger about.

But the biggest argument against silage was that it severely affected the quality of milk. "Feed silage and lose your market," was the cry heard around the country, particularly in Wisconsin. In some communities creameries refused milk and cream from farmers who fed their cows silage.

Even some of the farm magazines opposed silos. Wilmer Atkinson, publisher and editor of the *Farm Journal,* wrote in April 1881, "We shall not proclaim silage a humbug, because that may not be the right word to describe it. But it is only a nine day's wonder. Practical farmers won't adopt it, except one here and there, and in 10 years from now the silos being built will be used for storing potatoes, turnips, beets or ice."[10] Ten years later, Atkinson had changed his mind and was writing about the virtues of silos and encouraging farmers to get on the silage bandwagon.

## Silos Gain Popularity

The first silos were trench or pit silos, longer than they were deep. Many were built inside barns with the thought that less freezing would take place and groundwater would be less likely to find its way into them to spoil the silage.

Byron D. Halsted, in an 1893 barn-planning book, describes two such silos built by a pair of brothers named Buckly of Port Jarvis, New York. They constructed two pits under the cow-barn floor. The pits were 22 feet long, 9 feet wide, and 15.5 feet deep and were located side by side, with a two-foot wall between them. The pits were walled around and cemented watertight.[11]

**Corn for silage was cut into small pieces with a silo filler and then blown up a long pipe into the silo. Some farmers kept the silo filler pipes in place all year round, thus avoiding the sometimes difficult (or for some, scary) task of climbing to the top of the silo to install the pipe. Washington County, Highways 33 and 175 west of Allenton.**

The first silo at the University of Wisconsin's College of Agriculture was 27 feet by 12 feet by 15 feet deep. It was built in 1881. The walls were of rubble sandstone construction 18 inches thick. The inside was smoothed with cement. Because the silo was partially underground and the walls were of soft sandstone, the silage coming into the structure seemed to chill considerably because of the moisture the sandstone absorbed.

By 1887 the College of Agriculture was doing considerable research on silage and on the various varieties of corn suitable for silage. By 1889 the author of Wisconsin Agricultural Experiment Station Bulletin 29 estimated Wisconsin had about two thousand silos.[12] Most of these silos were constructed of wood and, although upright compared to trench silos, were rectangular.

**The earliest upright silos constructed in Wisconsin were square, not round. Kewaunee County, Sleepy Hollow Road and Gray Road.**

Spoilage was a great problem in these early aboveground silos because air got trapped in the corners and often leaked through the board sides, causing mold to develop. Providing enough weight on the silage when it was first put into the silo was also a problem. Farmers finally realized that by making their silos taller, the weight of the silage itself would serve to pack down the material. Thus silos became rectangles stood on end. A few of these rectangular silos may still be found in Wisconsin.

The taller silos created their own problems. No easy way to lift the corn fodder into the structure existed. The fodder was cut with a feed grinder, then pitched onto a platform, and then pitched once more into the open top of the silo. Spoilage continued in the tall rectangular and square silos because of the corners. The next step was to board up the corners with straight boards, making an octagonal silo. This led to a round silo, which became popular in the late 1800s and is the shape of most silos today.

## THE ROUND SILO

Professor F. H. King of the Wisconsin Agricultural Experiment Station is credited with developing the first successful round (cylindrical) silo. Even though the United States Department of Agriculture promoted the merits of the round silo in the early 1880s, few were constructed. Farmers thought their silos should be built inside their barns, and they felt round silos would be too difficult to build and repair there. Another reason farmers didn't immediately switch to the round silo was their basic skepticism of things round, as discussed in chapter 4. Farmers thought in terms of square corners and straight lines, not curves.

Professor King of the College of Agriculture introduced his round silo in 1891, and it became known as the King or Wisconsin silo. It had horizontal wooden siding with the interior wall separated from the exterior wall with an air space. With this development, the commercial silo industry began. Even though farmers didn't like the round structures, they could see readily that less silage spoiled in round silos than in ones built with corners. When silage spoiled, the farmer lost feed, and lost feed meant lost profit. This the farmer could understand. A basic dislike could be overcome if it meant increasing income, which was no small task for most dairy farmers.

Not only did the shape of the silos change, but the building materials changed as well. Early round silos were often constructed of fieldstone. The same stonemasons who constructed the fieldstone walls for the dairy barns springing up around the state could build silos as well. But it was a lot of work to build a fieldstone silo, and it took several weeks to complete the task. As labor costs increased, it became less feasible to erect a stone silo. The walls of the stone silo were often two feet or more thick. Even so, they did not resist well the great pressure developed by the stored silage, and they often cracked. In the late 1890s, lumber and other building materials gradually took the place of fieldstones for silo building. Walls became thinner, and strength was added by placing hoops around the round silos.

**Fieldstone silos required many hours and considerable skill to build. Waukesha County, Highway 67 and Country Side Lane.**

One farm often had silos of differing construction, usually built at different times. Here we see a wood-stave silo on the left and a brick silo on the right. Milwaukee County, circa 1917. **WHi Image ID 48052**

The wood-stave silo, popular throughout Wisconsin and the entire dairy region of the country, was built of 2-inch by 6-inch tongue-and-groove staves placed vertically and held together with steel hoops that were joined at the ends with iron lugs and nuts. This type of silo was relatively inexpensive, and it took only a day or two to put up. If properly cared for, wood-stave silos, particularly those made of redwood, lasted for years. Windstorms in the spring of the year, when the silos were empty, were the greatest enemies of wood-stave silos, for they toppled quite easily. Farmers learned how to attach guy wires to their silos, anchoring them to the ground and to their barns, thus lessening the chance of windstorms blowing them over.

But because the silo was usually located in the barnyard, where the cattle exercised, the cows often rubbed on the guy wires, loosening them and occasionally even tearing them up. The farmer sometimes ran into the guy wires with a piece of farm equipment and, because of the rushed summer schedule, put off repairing them. Thus the wooden silo was more vulnerable in a storm than it should have been.

Often the wood-stave silo was erected on a fieldstone wall, which had to be kept in good repair. Once cracks formed, water got into them, and during the winter the freezing and thawing widened the cracks and allowed air to enter the silo and spoil the silage.

Because the wooden staves expanded and contracted as they took on moisture and dried out, the farmer needed to adjust the hoops to keep the silo sturdy and airtight. Some silo hoops had stiff springs built into them, which allowed the silo to expand and contract automatically. Others had to be adjusted by hand.

Poured-concrete silos became popular in Wisconsin starting in the early 1900s. They were solid and built to last. Today, many remain standing long after other farmstead buildings have disappeared. Waushara County, County Road P, one-half mile east of 15th Road.

In the early 1900s poured-concrete silos came on the scene in large numbers. The Wisconsin Agricultural Experiment Station Bulletin 214, *Concrete Silo Construction* (1911), offered a step-by-step guide for building a reinforced concrete silo. When county agricultural agents began work in many Wisconsin counties after 1912, they had metal forms available to construct concrete silos and lent the forms to farmers for a minimal fee.[13]

Today, many poured-concrete silos may be found around the state. And another popular type of silo emerged that combined advantages of the wood-stave silo with those of the concrete silo. It is referred to as concrete stave and consists of concrete pieces. One manufacturer made them 10 inches wide, 30 inches long, and 2.5 inches thick. The concrete pieces were grooved, and the joints were sealed with cement plaster. Steel hoops, similar to those used on the wooden silos, held together the concrete-stave silos. The inside of the silos were usually coated with plaster or cement. Concrete-stave silos continue to be among the most popular silos constructed in Wisconsin.

Many other building materials were used for silos. Some farmers built their silos of concrete blocks 8 inches by 8 inches by 16 inches. With a concrete-block silo, however, it was difficult to ensure that all the joints were waterproof, and leakage meant spoilage. Tile-block silos also gained some prominence. The blocks were made of hard-burned or glazed tile especially molded to fit the curve of a silo wall. Good hard-burned bricks were also suitable for silo construction—if the cost of labor for laying them wasn't prohibitive. Brick silos were built with either single or double walls. The single wall four inches thick was the most common.

Metal silos, too, gave good service if properly cared for. It was necessary to keep the inside of metal silos painted with a refined tar, an asphaltic preparation, to prevent corrosion by silage acids. The metal silo was also most susceptible to wind damage when empty and thus required a good anchoring system. Another problem with a metal silo was that the silage froze quickly during cold weather. On the other hand, the metal silo was the fastest to thaw.

In the early days of silo use in Wisconsin, the farmer constructed his own silo, usually with the help of his neighbors. Later the farmer generally contracted with a silo manufacturer,

Tile silo under construction, circa 1900. WHi Image ID 37268

who usually came to the farm and erected the silo on the spot selected by the farmer. With the development and general acceptance of tower silos across the state, ensilage cutters and carriers soon were developed, making the task of silo filling considerably less difficult.

In the early 1900s Wisconsin took a census each year of the number of silos present on state farms. When the first such census was done in 1904, assessors reported 716 silos. In 1915 a census reported 55,992 tower silos in Wisconsin. About half were of wooden construction. By 1923 the number of silos reported in the state had grown to 100,060.[14] Today it is difficult to find a Wisconsin farmstead without one. It took about a quarter of a century for Wisconsin farmers to accept the silo, but once they did, nearly every farmer wanted one.

Corn was almost the sole crop put into silos until the second half of the 1900s. Then the University of Wisconsin's College of Agriculture began advocating grass silage, which was made from forage crops. Another round of controversy began, as farmers took sides either for or against grass silage. Some believed it was harmful to their cattle. Many used the same arguments used earlier against corn silage. Others thought it would completely replace hay. Cheese makers claimed grass silage spoiled the milk for cheese making. Today, partially dried grass silage has been accepted as an alternative to hay for cattle feed.

## METAL SILOS

The story of silos would not be complete without mentioning the big blue silos scattered throughout Wisconsin, the United States, and several foreign countries. Though metal silos gained some acceptance in the early days of silos in Wisconsin, they never were widely used. Silage froze too readily in them; they were susceptible to the ravages of rust, a process that the action of the silage acids accelerated; and they were easily toppled by a strong wind when empty.

Yet the A. O. Smith Company of Milwaukee, essentially by accident, got into the metal-silo business in the late 1940s. Back in the mid 1920s, the company developed a process for fusing glass to steel, resulting in a product that was both strong and resistant to rust. One of

In the late 1940s the A. O. Smith Company of Milwaukee began making glass-lined metal silos they called Harvestores. These were more expensive than more conventional poured-concrete and concrete-stave silos, yet many farmers bought them and applauded their superior qualities. One innovation was a mechanical unloader that worked from the bottom of the silo rather than the top. Green County, Dean Lane and Highway 60 near New Glarus.

the prime uses for the glass-fused-to-steel product was beer tanks, and in the Milwaukee area such tanks were widely accepted. In the 1930s A. O. Smith began manufacturing glass-lined water heaters, and they, too, became popular.

Looking around for even more uses for their glass-coated steel, Wesley G. Martin, a Smith engineer, learned from a farmer that one of the beer vats stood on end would make a good silo. That rather simple suggestion was the key to the birth of the big blue silos. Martin and his staff worked with E. B. Fred, then dean of the University of Wisconsin's College of Agriculture, to learn what problems farmers in the state faced in silage making. Dean Fred, with colleagues, had published a classic research paper on silage fermentations in 1925. Fred told the researchers that spoilage and the removal of frozen silage were two of the biggest problems. Another problem Fred mentioned was the pitting of concrete silo walls due to the action of the silage acids.

The Smith researchers, with the farmers' silage problems in mind, set out to develop a glass-lined steel silo that would solve these problems. They were not concerned about wall pitting with their glass-lined structure, but they were concerned about preventing feed loss through spoilage. They set out to construct a silo that was as airtight as possible.

The first silo they erected was a former beer vat that had been on display in the company's exhibit hall. They put it up on Swiss Town Farms, near Beloit. It was filled for the first time in the fall of 1945. Because the silo was nearly airtight, the researchers feared the silo walls would collapse due to the daily expansion and the nightly contraction, caused by temperature changes, of the gases inside. To combat this problem, engineers placed a big plastic bag in the structure. Expansion of gases caused by rising temperatures forced gas into the bag; contraction allowed the gases to escape back into the silo. This device equalized the pressure on the silo's walls without allowing outside air to come into contact with the feed once the silo had been filled and closed.

Another problem, rather obvious to be sure, was that if the silo was airtight, a person could not go inside to toss down silage, which was the traditional way of unloading silos. Smith came up with a mechanical unloader that worked at the bottom of the structure. Other equipment manufacturers had invented mechanical unloaders that worked at the top.

The first winter at Beloit proved successful. As expected, the glass-coated wall resisted attack from the silage acids. Though the silo walls were only one-quarter of an inch thick, the contents didn't freeze. The bottom unloader worked, but it was evident improvements were necessary. Coal-mining equipment provided the answer. Engineers looked at a machine that cut its way through beds of hard coal like a giant chain saw. The steel teeth were mounted on a chain-link pulley arrangement. Frozen silage and hard coal had some things in common, the engineers decided. They worked out an unloader that would gnaw its way through the silage, traveling 360 degrees around the structure. In 1949 the first one hundred production models of the Harvestore silo were manufactured. Why are the silos blue? The color is due to the cobalt in the glass mixture that is fused to the steel. Temperatures between 1,550 and 1,600 degrees F are used to fire the steel sheets and melt the glass so it fuses with the steel.[15]

By 1974 more than 46,000 Harvestore structures had been sold. During the farm crisis of the early 1980s, Harvestore silo sales plummeted from $140 million in 1979 to $21 million in 1984.[16] During this time and in the years following, farmers began constructing bunker silos, and many began using white plastic silage bags. Few farmers constructed upright silos; even fewer built the more expensive Harvestore silos.

## ✣ Field Guide to Wisconsin Silos ✣

**Wisconsin silos came in various shapes and sizes.** Early ones were square, later ones cylindrical. Building materials for silos included poured concrete, fieldstone, quarried rock, glazed tile, concrete staves, wood staves, brick, and metal.

Top left: An early square silo with a small, wood-stave round silo nearby. Top right: Concrete silos used for grain storage at a grain elevator. Center: Glazed tile silos. Bottom left: Silo constructed of quarried rock. Bottom right: Square silo and newer poured-concrete silo in back. Drawing by Allen Strang, courtesy of William Strang

Tucked in a valley surrounded by hills, this farmstead is typical of many found in southwestern Wisconsin. Iowa County, County Road A east of Hollandale.

# 11

# THE HEART OF THE FARM

## *The Farmstead*

A person could learn a lot about a farmer by looking at his buildings—at least that's what most country people thought. A neat, well-painted farmstead, with no trash lying around the yards, with barnyard fences in good repair, with grass and weeds trimmed, said, "Here lives a good farmer."

*Slack* was the word used to describe the farmer who didn't keep his buildings in good repair and who allowed the clutter from farming to accumulate around his farm buildings. No matter if the farmer's luck had gone sour and his income had all but stopped, he was expected to "keep things looking nice." An attractive farmstead reflected the farmer's outlook toward farming and, in a larger sense, illustrated his outlook toward life. Growing signs of clutter and disorder, with no attention paid to a loose board on the corncrib, a patch of wood shingles torn off the barn roof, a group of bull thistles growing along the barnyard fence, meant the farmer was losing hope. And lost hope was the first step toward a lost farm.

Besides saying something about the farmer's attitude toward his work, a farmstead described the nature of the farming practiced. From the turn of the century until World War II, most Wisconsin farms were diversified. Farmers raised several species of livestock; grew hay, corn, and oats to feed their livestock; and often raised a cash crop as well. Today, farmers tend to be more specialized.

## FARMSTEAD BUILDINGS

What buildings, then, would one find on a typical Wisconsin farm in the early 1900s? The house, barn, and silo were the primary structures. Occasionally there was a separate barn for horses, a granary, and a corncrib, and in the oldest farmsteads there may have been a barn that served solely as a storage place for hay and housed no livestock. Examples of this existed in the early German and Finnish farmsteads.

Most farmers had at least one hog house. Some of the oldest hog houses had once been log barns or even log houses. When the frame barn was built, the log structure often was given to the hogs. Hogs were hard on their shelters, and a full log building met the challenge of housing them well. Occasionally, farmers fashioned together a framework of poles, covered the sides with woven wire, boarded the roof, and blew threshed straw over the entire structure, burying it. The strawstack hog house, with its tunnel to the outside, was often the warmest shelter on the farm, for straw is an excellent insulator. (Strawstack shelters also housed heifers until they began milking, when they took their place in the dairy barn.)

Another popular hog house was about 12 feet long, 5 feet wide, and 6 feet high. The construction was A-frame; the roof served as the sides for the building. This

Two men survey a Wisconsin farmstead, complete with numerous buildings, pond, and windmill, date unknown. WHi Image ID 25818

hog house, constructed on skids so it could be moved from pasture to pasture, housed one sow and her litter. Larger hog houses could be found on many farms, some sheltering as many as one hundred hogs. Hogs had always been popular with Wisconsin farmers, especially since Wisconsin had become a dairy state. Although the price of pork rose and fell in roller-coaster fashion, farmers relied on hogs as "mortgage lifters."

Besides some type of shelter for hogs, there was often a shed for sheep and a shed for beef cattle. Sheep and beef cattle, however, were not as popular as hogs. Hogs and dairy cattle complemented each other well. But sheep, in addition to requiring housing, needed tighter fencing than dairy cattle. And because sheep had a tendency to eat pasture grass closer to the ground than cattle, many dairy farmers thought they'd kill good cow pasture if given the chance. Beef cattle also seemed to compete with dairy cattle. They ate basically the same types of feed as dairy cattle, required pasture, and needed shelter. And many dairy farmers claimed they just didn't understand beef cattle and therefore didn't want them around the farm.

A chicken house was found not far distant from the home on most Wisconsin farmsteads, for it was the women on the farm who cared for the poultry. Money from the eggs was used to buy groceries, or, in earlier days, the eggs themselves were traded for groceries at the general store in town. If the eggs were worth more than the needed supplies, storekeepers gave credit. If the supplies cost more than the eggs, the farmer made up the difference on his next visit. Some of the older Wisconsin farmers still mention going to town to "do their trading," though the practice ceased many years ago.

The brooder house served as a shelter for the baby chicks until they were old enough to be brought into the chicken house. It was usually a shed-roof building 12 feet to 14 feet long and 8 feet to 10 feet wide and was built on skids so it could be moved onto fresh grass each year.

Some farmsteads included hay sheds, particularly when the barn wasn't large enough to store sufficient hay. A hay shed often consisted of no more than a board roof supported by cedar posts braced to withstand windstorms. Some of the hay sheds were built square, with a four-sided hipped roof that came to a point and could be raised and lowered. As hay was fed from the shed, the roof was lowered, keeping rain and snow away from the top of the stack. (In many ways, the square hay shed was no more than a haystack with a roof.)

In a Finnish hay barn, the logs were purposely spaced apart so the wind could dry the hay stored there. Also, these barns were often built in the hayfields, away from the main barn. In case of fire, the farmer would still have his stored hay. Waukesha County, Old World Wisconsin, Eagle.

In the early days, farmers picked and husked corn by hand and stored it for winter use in cribs like this one. Waukesha County, Norwegian corncrib at Old World Wisconsin, Eagle.

Corncribs were a part of nearly every farmstead in Wisconsin, for almost all farmers grew at least a few acres of corn. These cribs took various shapes. The classic corncrib was a relatively long, narrow structure, wider at the top than at the bottom, consisting of boards placed a few inches apart so the air could easily circulate through the structure and dry the corn. To keep rats and mice out of the corn, the crib was usually built on concrete pillars that raised it at least two feet from the ground, and a metal shield was wrapped around the top of the pillars.

Nearly every farm had a granary, too. Sometimes it was simply a bin built into the haymow of the barn; the grain was fed by gravity to the livestock below. But often a granary was a separate building located away from the barn. Granaries attracted rats and mice, and farmers wanted to keep these pests away from their cattle barn if they could. Inside, the granary was divided into bins where oats, wheat, and rye could be stored. Sometimes there was a place on the second floor where ears of corn could be dried for planting the following spring, or where harnesses and other small pieces of farm equipment could be stored.

Machine sheds were often lean-tos attached to the barn, the granary, or some other farm building. Occasionally they were built as separate structures. In the pre-tractor days, most farmers had sheds where they kept their buggies and other horse-related equipment. Today, on many farms, these buildings have become machine sheds.

A pump house was built over the well. On many farms the pump house was also a milk house. The farmer cooled milk by placing the ten-gallon cans in a water tank and allowing well water to run through the tank. The overflow from the cooling tank drained into the barnyard tank, from which the livestock drank.

Other buildings found on the average Wisconsin farm included the outhouse, located a discreet distance from the back door of the house, and a smokehouse, often shaped like the outhouse and sometimes mistaken for the outhouse. (There was no place to sit in a smokehouse, and the lingering aroma of hickory smoke somehow didn't seem correct to the person who'd hurried into the smokehouse by mistake.) Homegrown meats, particularly hams and bacon, were commonly cured in farmers' smokehouses until after World War II, when most farmers were able to have electricity and thus access to refrigeration.

Most Wisconsin farmsteads included a woodshed, often a lean-to on the kitchen. Alternatively, the kitchen wing of the house might be divided into a kitchen and a woodshed, with a door connecting the two. Often a woodbox was attached to the kitchen wall near the kitchen range. An opening was cut in the wall so the woodbox could be filled directly from the woodshed. Occasionally the woodshed was built as a separate structure a few feet away from the back door of the kitchen. During the summer months, it served as a summer kitchen so the house might escape the heat generated by the kitchen range.

Each fall the woodshed was piled high with freshly split wood for the kitchen stove, for the dining-room and living-room heaters, and perhaps for the pumphouse stove as well. Once or twice during the winter, the farmer replenished the supply of wood in the woodshed. He cut the wood from the woodlot that was a part of nearly every farm, hauled the logs and limbs to the house with a team and bobsled, and then invited over the neighbors for a wood-sawing bee to saw the wood into stove lengths. Of course, when the wood sawers left, the larger

**Many farmsteads included buildings that served several purposes. In the foreground is what appears to be a corncrib on the left; the lean-to is likely a woodshed. Dane County, Highway 106 and Bingham Road.**

Tobacco sheds like this one were common in southern and southwestern Wisconsin during the years when tobacco was an important cash crop for dairy farmers. **WHi Image ID 30171**

pieces of wood had to be split into kitchen-stove-size pieces—a time-consuming but important task. Those pieces too knotty to split were burned in the heaters.

Depending on the part of the state in which the farm was located, other more specialized buildings could be found in the farmstead. In several southern counties and in several communities in western Wisconsin, particularly in Dane, Vernon, and Crawford Counties, most farmers had tobacco sheds, sometimes called tobacco barns. The first tobacco sheds were built after the Civil War, when the wheat crop failed several years in succession and farmers frantically searched for an alternative crop. Tobacco used for cigar wrappers was introduced to the state at that time, and it continues to be grown today, although in limited quantities. There was a close tie between tobacco growing and the Norwegians who settled in the state. Even today the bulk of Wisconsin's tobacco is grown in predominantly Norwegian communities. With antitobacco sentiment, tobacco growing in Wisconsin has declined rather dramatically. Where it was grown, it was almost always grown as a cash crop, to supplement a dairy operation.

Tobacco sheds could be easily confused with barns that housed livestock, until the observer looked a bit closer. One distinguishing characteristic of most tobacco sheds was the system of ventilation used to cure the tobacco leaves that were hung inside. Usually every other board on the side walls was hinged so it could be opened. Tobacco sheds were approximately 28 feet wide and varied in length depending on the amount of tobacco the farmer grew. Some were considerably longer than one hundred feet, but most were less than half that length. They were usually 15 feet to 18 feet high at the eaves and had a simple gable roof or, less often, a gambrel roof.[1]

Inside the tobacco shed, poles were placed four feet apart horizontally across the framing bents. The four-foot-long tobacco lath sticks, each supporting five to seven stalks of tobacco, were laid across the poles so the stalks and leaves hung downward. By arranging the poles at different levels in the shed, farmers could hang leaves from the peak of the roof all the way to the floor.

In late fall or early winter, the stalks were taken down during what the tobacco farmers called case weather. After the leaves were stripped from the stalks, the leaves were packed into boxes (cases) for shipping. Case weather was usually a few days of drizzle and fog that allowed the otherwise dry and brittle tobacco leaves to absorb moisture so they could be handled without shattering. Most tobacco sheds became machine sheds during the months of the year when they were not used for tobacco drying and storage.

Beginning in the late 1800s, farmers in central and north-central Wisconsin grew potatoes as a cash crop. For these farmers, potato money supplemented other farm enterprises, particularly dairying. Today, with modern technology, fewer farmers grow potatoes, but those who do often grow several hundred acres apiece.

Much early potato harvesting was handwork. Potatoes were planted by hand, weeded with a hoe, and dug with a six-tine fork. In the fall the farmer's children gathered potatoes into pails and dumped the potatoes into wooden one-bushel crates when the pails were full. At noon and at the end of the day, the farmer hauled the potatoes to the potato cellar, where

In the sand country of central Wisconsin, after wheat growing failed, potato growing became important. Potatoes were stored in potato cellars like this one. Waushara County, 15th Avenue north of Apache Drive.

# ✤ The Dooryard ✤

**A discussion of farmsteads would be incomplete without something about the dooryard,** a characteristic of rural farmsteads that is unknown to city dwellers, who speak of front yards, back yards, and sometimes side yards. The dooryard is an open area in the center of the farmstead surrounded by farm buildings and often shaded by oaks, maples, or great elm trees. The dooryard should not be confused with the barnyard, the fenced area adjacent to the barn where livestock exercise and are watered.

In the early days, before the coming of tractors and automobiles, the dooryard was where the neighbors tied their team when they came to visit. During haying season it was where the farmer halted the team while he and his children drew long, cool drinks of water from the well before unloading the hay.

A week or so before it was time to cut hay, harvest the oat crop, or cut corn, the farmer pulled his mower or binder under one of the big trees in the dooryard, greased the machine, repaired any broken parts not attended to since the previous harvest, and scraped off the dirt and grime from a year of storage.

The dooryard is the open area surrounded by farmstead buildings and often shaded by big trees. **WHi Image ID 59771**

There, in the shade of the big dooryard trees, a farmer repaired a broken machine after it ran against a stone; there he replaced a part worn out after many years of use.

During the threshing season, the neighbors who gathered to assist rested in the shady dooryard, after a heavy meal, talking about the oat crop and the summer weather. The dooryard was also where the neighbor boy courted the farmer's daughter, first in a buggy and later in a Model A coupe. Of course the farmer's bedroom looked out on the dooryard. If he wanted to check on what was happening in the buggy, he had only to part the curtains a wee bit and he knew.

The farm dog spent much of its time in the dooryard, resting on the porch off the kitchen or under one of the big trees. From this position, the dog greeted all visitors and was readily available when the farmer needed its assistance to fetch the cows to the barn.

The dooryard was also used for family reunions. Long tables were made from planks placed on sawhorses, and wooden folding chairs were borrowed from the church. Everyone brought something to eat, and the day was spent eating and sitting and talking under the shade trees.

In most farm homes, the front door opened into the parlor or the living room; the back door opened into the kitchen. It was the kitchen door that opened to the dooryard, and it was through this portal that everyone who knew country ways entered the house. It was easy to spot the city-bred salesman accustomed to they were stored until they were sold, usually in winter or early spring. Prices were almost always better several months after the potatoes were harvested.

These early potato cellars can still be found in many Wisconsin communities, particularly in Waushara, Waupaca, and Portage Counties. A typical potato cellar was located on the side of a hill, and part of the structure was above ground. If there was such a hill where the other farm buildings were located, this is where the potato cellar was constructed. If no suitable hill existed on the farmstead, the potato cellar was built where there was a hill. It was not uncommon to find potato cellars in the middle of fields or at the edge of woodlots—almost any place on the farm that could be reached easily with a team and wagon.

Potato cellars were not large; 20 feet by 30 feet was a common size. The walls of the potato cellar usually were about 10 feet high and constructed of fieldstones. On top of the cellar walls, the farmer constructed a wooden shed, usually with a gable roof. The shed had double doors on one end so farmers could drive wagons full of potatoes into the building. They dumped the potatoes through trapdoors in the floor to the potato bins below. A cellar of the size described here stored approximately three thousand bushels of potatoes.

The floor of the shed (the ceiling of the cellar) was constructed with two-inch-by-eight-inch joists covered with matched lumber flooring on both the upper and lower sides. Thus both the ceiling of the cellar and the floor of the shed above it were smooth. The open space between the floor joists was sometimes filled with sawdust and served as insulation.

The front of the cellar was exposed, providing an access door. Also in the front was a stove. The chimney jutted out the front wall. As the weather turned cold in late fall, the potato farmer started a fire in the cellar heater and kept it going night and day until he sold the potatoes. As a rule of thumb, the farmer hauled the potatoes to the potato warehouses in town only when the temperature was twenty degrees or above. If it was winter, he'd pile a bobsled high with sacks of potatoes and then cover them with blankets. If it was colder than twenty degrees, the potatoes would freeze on the sleigh even if they were covered. Of course the length of the trip to the warehouse also made a difference—the longer the trip, the more likely it was the potatoes would freeze.

## ARRANGEMENT OF THE FARMSTEAD

It is interesting how various farm buildings were arranged in relation to each other. Early settlers followed a few basic rules in placing their buildings. Sometimes these rules were based on ethnic considerations; certain ethnic groups tended to place their farm buildings in distinctive arrangements. Building placement was also dictated by natural features of the environment— a hillside, a woodlot, a stream, the direction of the prevailing winds. Usually, though, after all these factors were considered, where the buildings were located depended on what their function would be. And just as functional barns have considerable aesthetic appeal, so it is with farmsteads. The building arrangement that was most efficient and practical for the farmer usually was quite attractive as well.

Most decisions about building placement were common sense, yet some farmers failed to consider such obvious factors as where the spring meltwater accumulated and what happened after a heavy rain. J. H. Sanders, writing in 1893, had this to say about muddy barnyards:

> Nothing is more disgusting to the farmer's family, if they have any refinement about them, than a muddy, filthy barnyard; and the farmer's son who has been compelled throughout all his boyhood to wade through one of these disgraceful, stinking, wasteful cesspools is justified in his desire to leave the farm forever upon the first opportunity. Make the farm and all of its belongings attractive, if from no other motive than that of inspiring a love and respect for the farmer's calling in the minds of the boys who are born and brought up upon the farm.[2]

Water was one of the most important considerations in determining the location of a farmstead. Many farmsteads were near water—a stream or river, a spring, a lake. When such natural, readily available sources were not present, the farmer considered whether he could profitably dig a well for water or could haul water from a nearby source.

Well digging was no small task, particularly in those parts of Wisconsin where the digger had to descend more than one hundred feet into the ground. Yet that is exactly what many farmers did. They rigged up a rope-and-bucket affair, hitched an ox to the end of the rope, and

Farmers paid particular attention to how they arranged the buildings in a farmstead. The barn, hog house, and chicken house were usually placed downwind from the farmhouse. The buildings were also clustered by function. For instance, the corncrib was located convenient to the hog house. Buffalo County, Highway 35 and Highway 25 north.

then slowly dug their way to water, cribbing up the sides of the well as they dug ever deeper. These early wells were dug to water because modern well-drilling equipment was unknown. In some instances farmers assumed they could dig a well without difficulty and proceeded to erect their farm buildings. Later, to their dismay, they discovered the amount of digging was much greater than they had expected, or they struck huge boulders or even bedrock, which prevented them from digging to water.

Such was the case with one farmer who built a farmstead in central Wisconsin. The house, complete with fieldstone walls and cellar, had been erected, and the barn and other outbuildings were up. While the building was going on, the farmer hauled water from a stream more than three-quarters of a mile away, fully confident that he could easily dig a well near his house. When he began digging his well, he discovered, after many attempts, that the task was impossible. Time and time again he struck huge boulders that got in the way of his progress. He also found he needed to dig much deeper than he had anticipated.

Finally, he quit digging and continued hauling water from the creek until he could decide what to do. In desperation, he sought out another building site on his farm, one where he could dig a well. The decision was difficult, but it was the only one that seemed feasible under the circumstances. He found a likely place for a well half a mile from his buildings. There he dug to water. Then he commenced moving his newly constructed buildings to the well site. This proved both expensive and time consuming, but a constant supply of water was essential to his farming.

Besides drainage and the availability of water, farmers considered the prevailing winds when they selected sites for their farmsteads and decided on the placement of buildings. Because winter winds blow primarily from the west and northwest, farmers located their buildings on the east side of a hill or on the east or south side of a woodlot to avoid the frigid blasts. Farmers also considered the summer breezes and the natural smells associated with a cow barn, chicken house, and hogpen. They tried to locate their homes to the south or

**The danger of fire was always a consideration when locating farm buildings. The buildings were placed close enough for convenience but not so close that a fire in one of them would destroy them all. Door County, West Meadow Road and Highway 57.**

southwest of the other buildings, as the prevailing summer winds blow from these directions. Early agricultural publications gave specific instructions about arranging the buildings in the farmstead. A 1922 book said:

> Sunlight is one of the most potent enemies of dirt and disease. Germs do not thrive in sunlight, and dirt is more readily detected in a bright, cheery room than in one that is dark and dreary. It is very desirable, therefore, that all shelters of human and animal life receive the utmost benefit of the sun's rays during the winter and of the cooling breezes in the summer time. The principle of orientation is the arranging of the various parts of the building so that this end may be attained.[3]

The book suggested the farmer should select a south or southeast slope for the farmstead. The barn for livestock should be placed with the long axis north and south "so as to give either side the benefit of the sun for one-half day, and also to allow cooling summer breezes to blow through the buildings."[4]

An 1893 plan book recommended that farmers locate farm buildings in reference to the fields and pastures to save much valuable time driving cattle to and from the barn. According to this advice, the ideal location for the farm buildings is as near as possible to the center of the farm. Many farmers followed this advice, as evidenced by the long drives that lead to numerous farmsteads. To place their farmsteads in the center of their farm, some farmers had to build drives half a mile long or longer.

Most farm women, who often had fewer social contacts than their husbands, didn't much like this arrangement. They wanted to live near the main road so they could see their neighbors when they drove by. Thus it is not unusual to find many farmsteads cut in two by a road, with the barn and outbuildings on one side and the house on the other. During the days of horses and buggies this was quite a useful arrangement, for the farmer and his family could have easy, continuing contact with all travelers passing by. But with the advent of the automobile and hard-surfaced roads, travelers raced by the farm without even waving at the farmer waiting to cross from the house to the barn. The problem became worse when many country roads were widened to accommodate even more traffic. Where once buggies leisurely passed by the farmstead, now diesel trucks roar by day and night, rattling the dishes and interrupting sleep.

As the years passed and the farm operation grew or changed emphasis, additional buildings appeared on the basic farmstead. Their placement was determined by such practical considerations as available space and purpose.

The barn was the centerpiece of a farmstead. Barns were the largest buildings in the cluster of farmstead buildings and generally the first seen by passing motorists. The farmhouse was of secondary importance. Fond du Lac County, Memorial Drive northeast of Lindsay Drive.

A farmstead like this one (photographed in Wisconsin circa 1873) was a showplace for a farmer, a place to display how well the farm operation was doing by building attractive fences, keeping buildings painted and in good repair, and maintaining neat and tidy grounds. **WHi Image ID 25533**

With early farmsteads, the arrangement of buildings showed ethnic influences, according to some researchers. When he was working with the Wisconsin Historical Society, Gary Payne researched how buildings were located in early German farmsteads. The basic arrangement is known as a *Vierkanthof,* or four-sided-yard plan. The major buildings of the farmstead, traced to farms in the northern part of Germany, were located on the sides of an open square.[5]

The house was typically on the south side of the square, where it could benefit from the sun during the long winter months. When it didn't face south, it faced some important landscape feature such as a river or road. The cow barn was built on the north side of the square. It usually was a building of one and one-half stories, with cow stalls on the lower level and feed storage above. Another two-bay barn was located on the west side of the square, with a threshing floor oriented east and west. The north bay of this barn was used to store unthreshed grain; the south bay was used to store hay. The threshing floor had large doors opening on both the east and west sides, which allowed the wind to pass through, making hand threshing easier.

The east side of the square typically included a horse barn or a building to house sheep, pigs, or poultry. In this plan the well was found between the cow barn and the house so it could serve the needs of both the farm family and the livestock.

Continuous architecture was another approach to farmstead arrangement. Though rarely found in Wisconsin, it was popular in several New England states, particularly in Maine.[6] As the name *continuous architecture* implies, all the buildings in the farmstead were connected. Usually the buildings were arranged in a straight line, but sometimes they were arranged in an L. The house was usually on one end of the farmstead. Connected to the house was a wood-shed, then a wagon shed and the main barn, followed by a hog house, horse barn, or sheep barn, depending on what type of livestock the farmer raised. In this arrangement, the farmer could walk from the house to any other building without going outside. During the bitterly cold and snowy New England winters, this was a most convenient arrangement. The problem with this arrangement was the danger of fire. In his *An Age of Barns,* writer Eric Sloane points out that continuous architecture was banned in much of New England in the 1600s because of fire hazards, but there is no record of anyone ever paying a fine, and in the 1700s the ban was lifted. Even today it is easy to find examples of continuous architecture in Vermont, New Hampshire, and Maine.

Farmsteads continue to be popular subjects for artists and photographers. The blending of shapes and colors and the contrast of the buildings with the natural environment provide attractive, appealing scenes. The number of farmsteads, though, is shrinking, which is the subject of the next chapter.

With a damaged roof, it is only a matter of time before this barn falls. Dodge County, East Salem Road near Columbus.

# 12

# Vanishing Landmarks

## *Barn Preservation*

Wisconsin is losing its old barns. In 1935 the state had 200,000 farms. If every farm had at least one barn, Wisconsin would have had 200,000 barns in 1935. By 2008 Wisconsin had 78,000 farms and only 13,110 dairy farms. Of course, many of these farms still have barns, but especially on dairy farms many of the barns are of new, more modern construction.[1] Thousands of old barns have disappeared—exactly how many is not known. In 2008 University of Wisconsin researchers found that about 80 percent of the dairy herds in Wisconsin averaged one hundred cows or fewer, and the majority of these farmers were milking and housing their cattle in the traditional stall barns built in the early 1900s. Many of these old barns had additions as dairy herds increased in size from the 1940s through the 1980s.[2]

Extrapolating from these data, approximately seven thousand dairy farmers are using their old barns to house their dairy cows, a fraction of the original number of barns. Another unknown number of these barns are used for other agricultural purposes—livestock housing and feed storage. And another substantial number are used for what is called "adaptive reuse."

## The Disappearance of Barns

What's happened to thousands of old barns? Many disappeared during tough economic times. Immigrants who moved into the cutover areas of northern Wisconsin in the late 1800s and early 1900s found themselves on marginal agricultural land. As long as farm prices were high, they could make a go of it, but as soon as prices for their produce dropped, they had insufficient margin to continue. This was particularly true during the Depression years of the 1930s. Farmers left their farms, sometimes selling what they could at auction but too often simply abandoning their land because they couldn't pay the taxes. Many of these farms became county property, and the buildings were allowed to fall down and decay after their occupants left.

World War II provided a boom for most Wisconsin farmers, and those who had made it through the Depression could now exist quite comfortably on marginal land. This was the case

The remains of an early log barn. Sheboygan County, Division Road near Kettle Moraine State Forest Northern Unit.

for many farmers who had settled on the sandy soil of central Wisconsin, where I grew up. But after the war, when farm prices declined once more, an economic squeeze prevented farmers from making a satisfactory living, and again farms were abandoned. These shifts were not as dramatic as those that occurred during the Depression, however. From the late 1940s into the 1960s, boys and girls who had grown up on the land left the farm after finishing high school to find factory jobs in Milwaukee, Racine, Kenosha, Janesville, Oshkosh, Green Bay, Chicago, Minneapolis, St. Paul, and other cities. I left the home farm in the 1950s; so did my two brothers.

During those years, and continuing somewhat to today, when farmers reached retirement age, they sold their farms to neighbors who were expanding or to land developers who created everything from Christmas tree plantations to campgrounds and other recreational retreats.

During the 1950s to the present time, thousands of farmers moved off the land. Some barns were purposely burned to take them off the tax rolls. Some were torn down and the lumber used for other purposes. Many of the barns were abandoned. Traveling through the central and northern parts of the state one can easily find these old barns, most of them decaying.

During the past several decades, Wisconsin has seen an increase in farm sales and lost barns. For many years, Wisconsin was the leading producer of milk in the United States, but no longer. California has now wrested that title away. Wisconsin had 2.1 million dairy cows in 1960. By 2008 that number had fallen to 1.25 million.[3]

For those farmers still milking cows, other changes have affected the old barns. When I was growing up on a dairy farm in Waushara County in the 1940s and 1950s, we milked ten cows. During the 1940s we milked by hand; it was hard work, and it took lots of time. After World War II my father bought a Sears, Roebuck Riteway milking machine that was powered by a Briggs and Stratton engine because we didn't yet have electricity. Now we could milk fifteen or twenty cows in the same amount of time and with a lot less hand labor. Twenty milking cows was a rather common size for a family-run dairy farm in those days following World War II.

By 1960 the average herd size was still only twenty-one cows. By 2008 the average herd had increased to ninety-one cows. Two-hundred fifty dairy herds in the state had five hundred cows or more.[4]

With larger dairy herds the old barns became too small and too inconvenient for the dairy farmer trying to eke out a living milking cows and competing with the likes of California, Arizona, and other western states where dairy farming has increased in importance during the past few decades. These larger dairy farms, many of them still family operations but some corporate endeavors, built modern barns equipped with the latest technology for housing dairy cattle. The old barns, many built around the turn of the century, were forgotten, becoming mere historical punctuation marks in the story of Wisconsin agriculture.

**This small old barn is leaning and will soon collapse, leaving behind a silo to mark its memory. Columbia County, Ziehmke Road.**

In addition, even if modern-day farmers wanted to build new barns following the style of the historic barns, they couldn't afford it. Labor costs are too high to allow a crew of eight or ten carpenters to work several months sawing, drilling, and pegging together the timbers of a barn. The work must be done in far less time. And it is. Today's "new" barns bear little resemblance to the old two-story bank barns found throughout the state. These modern-day barns are one-story affairs—metal construction with cloth sidewalls that can be rolled up during the hot days of summer and lowered in winter. Many of them house more than a hundred milk cows each.

Over the years, old barns have been lost to lightning, windstorms, and fire. Because they are so large and usually stand up higher than any other building in the farmstead, barns are natural targets for both lightning and wind. Lightning rods help prevent some, but not all, lightning fires. And almost nothing can be done to protect a barn from a windstorm, especially in the spring of the year, when the haymows are empty.

**The barns stand, but the shed in the foreground collapsed under heavy winter snows. Such disasters often happen after a building's roof begins to deteriorate. Jefferson County, Klement Road and Highway 106.**

I will never forget the year the barn on our home farm was blown off its foundation by a fierce straight wind that continued nonstop for nearly twenty-four hours. Part of the barn wall toppled, burying calves, injuring cows, and creating a scene seldom captured in even the most vivid horror movies. One windstorm changed our farming operations forever.

Fire is one of a barn's worst natural enemies, for once a fire starts it is almost impossible to save the structure from total destruction. Spontaneous combustion of hay put into the barn too green is a cause of barn fires that goes back to the earliest times and continues today. Hay improperly cured becomes hotter and hotter until it bursts into flame. Often the fire will smolder for many days in the heart of a haymow before it breaks through to the surface and rapidly consumes the entire barn. Many mysterious barn fires have been caused by spontaneous combustion. Barn fires result from other circumstances as well—careless smoking, short-circuiting of electrical wires, and lightning.

Whatever the cause, a barn fire was a major catastrophe for a farmer. He mourned the loss of his barn as he would mourn the loss of a member of his family. There was a barn fire in my neighborhood when I was a youngster. Several short rings jangled on the party-line phones in the area—a general ring, it was called—and the telephone operator announced that a neighbor's barn was burning. Farmers gathered milk pails, milk cans, and shovels and rushed to the neighbor's farm. By the time the neighbors began arriving, flames were shooting high in the air, spewing out of the haymow windows and sending a huge cloud of black smoke into the night sky.

The farmer shouted that he couldn't remove several cows from the barn. We heard the bellowing of the helpless animals as they screamed to be released from a certain death in the barn's stable, but there was little the crew of neighbors could do except pour water on the adjacent buildings to prevent the fire from spreading and console the farmer and his wife who were witnessing much of their life's work being consumed by the yellow flames.

In a short time the upper framework of the barn collapsed, sending showers of sparks over the farmstead. The neighbors quickly poured water on the sparks and pounded with shovels and water-soaked burlap bags. The fire was contained, but the barn was lost, along with several head of livestock.

Because many tons of hay were stored in the barn, the fire continued throughout the night and most of the next day. A small crew of neighbors took turns watching to make certain other buildings were not ignited. It took a week for the ashes to cool enough so the bloated cattle could be pulled from the rubble and buried. At times the smell of burned and decaying animals was almost overpowering to those who helped with the terrible task of cleanup. The intensity of the fire's heat was evident everywhere. Metal stanchions had melted and twisted into strange shapes. Glass from the windows had not only shattered but often melted into grotesque globs. The farmer and his family went through the motions of life as usual, but it was obvious they were grieving, and understandably so, for they had lost much in the fire.

**One of a farmer's worst fears is a barn fire. Once such a fire begins, it is almost impossible to extinguish. WHi Image ID 23776**

Around the state can be found the remains of burned barns—a fieldstone stable wall, sometimes a silo. For some farmers the loss of a barn meant giving up farming and moving away to make a living at something else. This was particularly true in some of the more marginal farming areas in central and northern Wisconsin.

Barns are disappearing for other reasons, too. Near any large city, housing developments, shopping centers, and industrial buildings are replacing barns. Rather than grow upward, as New York and some other cities do, Wisconsin cities expand outward into the countryside. Housing developments have sprung up on once-rich farmland like mushrooms after a spring rain. Many of these new home owners want a spot of land, perhaps only a quarter acre, for their children to play and their gardens to grow. Some of these new city dwellers grew up on farms and simply are uncomfortable living in crowded condos or apartment buildings. Others have built in the country simply because they could, because it seemed there was an unending amount of available land. Suburban development engulfs barn after barn in Wisconsin. One summer a farm grows alfalfa and the barn houses Holsteins; the next summer the same fields grow a crop of new homes and condos and the barn succumbs to a wrecking crew. The cities sprawl farther and farther into the countryside.

In addition to suburban housing developments, shopping centers are savage enemies of farmland and barns. As cities became larger and more congested, and the number of automobiles increased each year, people discovered there was no place to park their cars. How can a person shop if there is no place to park? Shopping-center developers quickly saw what was happening and began building entire shopping districts in the country, their biggest selling point to the consumer being free parking. Of course all of the farm buildings had to be removed before the land could be "formed" to fit the requirements of the shopping center.

There are yet other reasons why Wisconsin's barns are disappearing. One argument goes, "There are so many barns in this state, what difference does it make if we lose a few here and there?" What is abundant, we take for granted, never realizing that what is common today may be rare tomorrow. The story of the end of the passenger pigeon makes the point. There was a time when the air was filled with passenger pigeons; they were everywhere. They were common. And then there were none. Not even one.

An underlying reason why our historic barns are disappearing is the speed with which Americans brand something obsolete—the old barns are often described this way. We value highly that which is new and modern. New structures, whether they are houses, office buildings, industrial buildings, or barns, are viewed as status symbols.

Traveling around Wisconsin, it is possible to find many abandoned barns, forgotten structures. Old barns don't die easily, though, particularly if their roofs remain intact. The paint will fade, and the boards will take on a shade of soft gray, but even with no maintenance at all, the barn will stand. Once the roof begins leaking, however, either because the wooden shingles have begun to rot or because they've been partially blown away by a windstorm, the barn's days are numbered. The rain seeps in and the roof boards begin to rot. The rot moves through the structure, slowly during the first years, more rapidly as increasing amounts of moisture enter the doomed building. During a heavy snow some winter the roof collapses onto the mow floors, leaving only the timbers standing. It is only a matter of time before the timbers will topple and eventually rot if some scavenger doesn't pull them out of the barn and sell them.

A once-proud barn with an attached lean-to, now in bad repair, is on its way toward total collapse. Door County, Highway 57 and Grove Road.

An abandoned barn is marked for death. It will die quietly, surrounded by field crops, a new housing development, a shopping center. No one notices except the occasional person who drives by and exclaims, "What a terrible eyesore! Why doesn't someone tear down that old building?"

Sometimes there is no sign or almost no sign that a farmstead even existed at a particular location. Lilac bushes are one of the clues to look for. All of the farm buildings may have disappeared, but unless the lilac bushes are purposely removed, they endure, coming forth with aromatic lavender flowers each spring, reminding all that a farmstead once stood there.

More often these days, the farmstead is removed entirely, including the lilac bushes. Near the farm where I grew up in Waushara County stood a very attractive farmstead with a large white barn, a substantial house, and a number of outbuildings. One day I heard the farm had been sold, as had several other farms in the neighborhood. The report was that the farmer who bought the land planned a large potato-growing operation. A few months later I drove by the farmstead, and I could not see one sign of it. Where there had once been a barn and a house and other outbuildings was now a potato field, with a huge irrigation sweep swishing its way across the place.

## WHY SAVE THE OLD BARNS?

Historic structures are a link with our past. They provide tangible information about the culture and technology of an earlier time. Wisconsin was a rural state; in many ways it still is.

Whether we realize it or not, we all have a tie to the land. Some of us grew up as farmers. Many more of us had parents or grandparents who were farmers.

Barns are symbols of the land, a tie to the natural world, a reminder that we are all a part of the environment, even when we try hard to be apart from it. Barns remind us of our historic bonds to the land. They help us trace our roots and, in turn, assist us in knowing who we are. We are all people of the land, dependent on it and nourished by our relationship to it. The old barns contain stories of several generations, memories of making hay and caring for cattle, of rainy days and the drum of raindrops on the barn roof, of frigid days with frosty windows and intricate frost-covered cobwebs in the haymow.

Barns tell us of our agricultural heritage and help us better understand the nature of farming during an earlier time. The old farm buildings reveal the technology of the time—how the old buildings were constructed and the uses for which they were designed. The old barns help us trace the history of agriculture in Wisconsin, from a time when we were predominantly a wheat-growing state to when we, slowly at first and then more rapidly, moved toward becoming the leading dairy state in the nation.

Barns are tangible reminders of our ethnic roots—the German half-timbered construction, the Swedish log barn, the Finnish hewn logs, the English town barn, the Swiss pentroof design,

**In this abandoned, two-story, arched-roof barn, the roofline remains true, one feature to consider when deciding if a barn is worth saving. Washburn County, Highway 53, Trego.**

Norwegian tobacco barns, Icelandic fish barns, and the work of the Welsh stonemasons who constructed elaborate fieldstone walls.

From an aesthetic perspective, many Wisconsin barns are truly works of art and should be preserved for that reason alone. Whereas modern-day farm structures have a sameness about them, the old barns are distinctive. Each one has a character of its own and makes a unique aesthetic contribution to society. Beyond these individual artistic qualities, the old barns contribute to the overall beauty of the countryside. Destroy an old barn, and the beauty of a landscape may be destroyed as well.

Ruth Olson, associate director for the Center for the Study of Upper Midwestern Cultures, University of Wisconsin–Madison, has said this about the importance of preserving old barns: "Like other historical artifacts, barns reveal our occupational and ethnic histories, as well as cultural patterns. Because of their size, the complexity of their detail and that they continue as working buildings (often in different roles) barns can make history come alive."[5]

We are not talking about preserving the old barns to benefit just a few. The old barns are of the people, and their preservation is for the people, all of the people. Everyone has roots that trace back to the land. And the old barns are of the land.

From a practical perspective, barns can be used for a variety of purposes. Most of these old buildings were well built and viewed by many as too good to take down. The challenge,

An old round barn with a story to tell. Ozaukee County, Grafton, date unknown. WHi Image ID 2054

then, is not only to save them but to find a use for them. Saving Wisconsin barns does not mean promoting thousands of barn museums scattered around the state. Buildings are meant to be serviceable, whether they are old or new. The old barns are no exception.

## Restoring Barns

Some people assume that renovating an old structure is complicated and expensive and will result in a building that is more costly to maintain. In fact, if the building is basically sound, rehabilitating it can be less expensive than new construction.

Sometimes disassembling an old barn is the best way to save the building's posts, beams, and boards. Chippewa County, Highways 64 and 53.

Useful questions to ask when considering saving an old barn are: Is this old barn in good enough shape to save? What are possible new uses? What will it cost to rehabilitate it? Is this a practical thing to do? A quick check of the following can tell you if your barn is suitable for rehabilitation.

- **Framing.** Are the posts, beams, sills, rafters, trusses, and joists free of rot?
- **Foundation.** Is it free of cracks? Is it settling and shifting? Is the mortar crumbling and falling away?
- **Roof.** Are there water stains or evidence of rot on the beams and roof boards? Are sections of roof covering missing?

- **Exterior walls and roofline.** Does the barn appear straight? Or is there a bow in the roof? Are the walls sagged or pulled out?
- **Interior.** Are there any rotted boards in the existing floors?
- **Location.** Is the barn conveniently placed with good access to other buildings and the farmyard?
- **Utilities.** What is the condition of the wiring and plumbing?[6]

## Barn Preservation Organizations and Resources

For those interested in preserving and rehabilitating barns, there are a number of state and national organizations offering information and technical assistance.

### Wisconsin Barn Preservation Initiative

In the fall of 1993, a small group of people met at the Wisconsin Historical Society's headquarters in Madison. Along with members of the historical society, participants included representatives from the Wisconsin Trust for Historic Preservation, University of Wisconsin–Extension, and several interested individuals. The purpose of the meeting was to see if there was interest in starting a save-the-barns project. Similar efforts had been successful in other states, particularly in Michigan. Bill Kimball of the Michigan State Cooperative Extension Service explained the project in his state and described how hundreds of people had turned out for workshop meetings. Following the exploratory meeting in 1993, a public meeting was held in Waupaca in April 1994 to an overflowing meeting room.

Currently a furniture and gift shop, this well-cared-for old dairy barn will stand for many years. Waukesha County, Highway 83 and Interstate 94.

Since that first meeting, numerous barn preservation workshops have been held throughout the state, attended by hundreds of people interested in old barns and how to save them. Charles Law, University of Wisconsin–Extension, was one of the founders of Wisconsin's barn preservation program. When asked about the future for barn preservation work, Law said, "The interest will continue at a high level with land use changes. For increasing numbers of people, old buildings are an asset to a piece of property. They keep the rural landscape alive."[7]

The University of Wisconsin–Extension also offers extensive materials on barn preservation at its Web site, including estimated costs for barn repair and remodeling and a list of contractors, consultants, and architects interested in saving Wisconsin's barns. For more information on the Wisconsin barn preservation program, visit www.uwex.edu/lgc/barns.

## Wisconsin's Historic Preservation Program

The Wisconsin Historical Society's Division of Historic Preservation administers the program that promotes the preservation, protection, and use of Wisconsin's irreplaceable prehistoric and historic properties, including barns.

Established in 1989, Wisconsin's State Register of Historic Places lists state resources that are important because of their historical, architectural, engineering, archaeological, or cultural significance. Any person may nominate a property to the State Register by submitting a nomination form to the Division of Historic Preservation. A Historic Preservation Review Board reviews applications and makes decisions.

There is also a National Register of Historic Places (NRHP), administered by the National Park Service. The National Register contains thousands of Wisconsin entries.

Both Wisconsin and the federal government offer investment tax credits for rehabilitating historic buildings. For details, visit www.wisconsinhistory.org, or contact the Division of Historic Preservation, Wisconsin Historical Society, 816 State Street, Madison, WI 53706. The Wisconsin Historical Society has an extensive collection of barn preservation materials available on its Web site: www.wisconsinhistory.org/hp/barns.

## BARN AGAIN!

BARN AGAIN!, a program of the National Trust for Historic Preservation in cooperation with *Successful Farming* magazine, launched in 1987. Its purpose is to promote the preservation and practical use of older farm buildings. Since its modest beginning, BARN AGAIN! has become a national resource for thousands of people concerned about the future of America's rural heritage. The BARN AGAIN! program has documented hundreds of examples of how barns and other farm buildings have been rehabilitated for productive new uses, from housing dairy cattle or storing machinery to becoming stores, homes, studios, and museums. Annually, the organization provides awards drawing national attention to farmers and ranchers who have preserved their buildings.

For further information about BARN AGAIN!, including a list of publications providing practical information for restoring historic barns, visit www.agriculture.com/barnagain/index.html.

### National Barn Alliance

The National Barn Alliance (NBA) began as an informal network of cooperative extension faculty, historic preservationists, and people interested in barn preservation. Initially Illinois, Indiana, Michigan, Ohio, and Wisconsin land-grant universities were involved. The organization has expanded to include other states, state historic preservation offices, statewide barn preservation groups, and national organizations such as the National Trust for Historic Preservation. NBA collects detailed information about the history, physical characteristics, and unique features of barns and rural historic structures and works for their long-term protection. For more information visit www.barnalliance.org.

## CONVERTED BARNS

Many barns have been converted for purposes other than housing dairy cattle. Some old barns have become storage places for boats, travel trailers, and automobiles. They have also become warehouses for lumber, fertilizer, beer—nearly anything that requires space under cover.

The large, open spaces of a barn's interior lend themselves well to restaurants and banquet halls. At Old World Wisconsin Historic Site in Eagle, one of the octagonal barns originally built by Ernest Clausing in Ozaukee County serves as a restaurant, one of only two or three Clausing barns still standing (albeit not on its original site).

The Round Barn Lodge in Spring Green is another unusual barn now being used as a restaurant. Built in 1914 by Ewalt Andreas, it was used as a dairy barn until 1949. The barn housed between twenty and thirty cattle and was built for a cost of between eighteen hundred and twenty-five hundred dollars. As in many round barns, the silo was located in the center, making the task of cattle feeding easier. The barn was converted into a dining establishment in 1949 and was renovated into a full-service restaurant and bar in 1952. A two-story lodge was added in 1974, and more rooms were added in 1988.

Located between Madison and Verona, Quivey's Grove features an 1855 stable and a mansion, both of which are today popular dining spots. The restaurant and the Stable Tap opened

Once a log barn, now an antique and gift shop. Door County, 10079 Highway 57 north of Bailey's Harbor.

in 1980. Both buildings are now listed in the National Register of Historic Places. A new entryway and main bar room were added to the stable in 1989. The building materials for the remodeled stable were salvaged from nineteenth-century barns. Diners in the stable enjoy a close-up look at hand-hewn timbers that were sanded and sealed but not stained and original stone walls that are twenty inches thick. The hand-hewn timbers on the first floor of the stable are original and of several types of wood, including walnut. A photograph of a barn stall near the entrance reminds people that the stable housed horses as late as May of 1979.[8]

**This round dairy barn became a restaurant many years ago. Sauk County, Highways 14 and 60, Spring Green.**

This former stable is now the Stable Tap at
Quivey's Grove, a popular Madison-area restaurant. Dane County, 6261 Nesbitt Road, Fitchburg.

A former dairy barn now serves as headquarters for Eplegaarden, a popular orchard. Dane County, 2227 Fitchburg Road, Fitchburg.

An old gambrel-roof dairy barn serves as the retail store at Laverne and Betty Forest's Epelgaarden orchard in Fitchburg. The barn's former haymow now hosts a variety of activities, from wedding receptions to a Halloween haunted house.

As farm families have found creative ways to diversify their operations, many of them have converted barns and other farm buildings to new uses in the process. At Schuster's Playtime Farm near Deerfield, Wisconsin, the family has been restoring their centennial round barn as the centerpiece of their pumpkin farm business. They also rent out the barn for functions enhanced by a rural atmosphere and lots of space.

Old barns star in the operations at a number of Wisconsin wineries, including the Apple Barn Winery in Elkhorn and Botham Vineyards and Winery in Barneveld. And the Big Barn Workshop and Kitchen Shoppe near Eagle River was once a large dairy barn.

It is not unusual for barns to be converted into living space. The Schwegel family of rural Mazomanie recently took on the onerous task of turning a 1910 post-and-beam dairy barn into a home. When they started the renovation, the barn was dilapidated, with one wall near collapse. The result of their rehabilitation is an attractive and highly livable home.[9]

Indeed, those interested in converting a barn into a home face several significant challenges. Besides the obvious work of repairing roofs, restoring walls, creating private spaces, and installing plumbing and fixtures, the barn-turned-house results in an enormous amount of space that will require heating in the winter. Maintaining the integrity of the old barn and its beauty while making the structure practical and affordable to live in is a daunting endeavor, to say the least.

## ⚜ Town of Chase Barn ⚜

**In Oconto County stands a barn built entirely of fieldstones from the area.** Designed and built by Daniel E. Krause in 1903, the barn is 100 feet long and 60 feet wide, with post-and-beam framing and stone walls two feet thick. The building boasts many unique features, including large arched doorways that allowed for wagon loads of loose hay to be pulled into the barn and unloaded into the haymows.

By the 1990s the barn had suffered from years of neglect. The north wall had begun to lean outward, causing many cracks in the surrounding walls. In 1995 then-owners Stanley and Casmir Frysh began a restoration of the barn, which included straightening the leaning stone wall. A later owner got the barn listed on both the National and State Registers of Historic Places. Realizing how important the barn was to the region and to Wisconsin, the Town of Chase purchased the barn and ten acres of land in 2007 and continued the restoration. The Chase Stone Barn Committee is developing a master plan for the barn's continued restoration and as of this writing is sponsoring public events at the barn and other efforts to raise money for the project. The project has also received a challenge grant from the Jeffris Family Foundation of Janesville.

In recognizing the importance of an old barn and making it a priority to restore it, the town of Chase community is a model for other barn restoration projects.

Built entirely of fieldstone from top to bottom, this unique barn is now owned by the Town of Chase and is being restored. Oconto County, County Road S and Schwartz Road, Chase Township.

Because of their sturdy construction, old barns are good candidates for other uses. But beyond the practical, old barns evoke feelings of warmth and nostalgia in many people. These feelings, in part, have motivated people to convert old barns into recreation halls, taverns, motels, antique stores, and homes, both to reuse their soaring spaces and to keep them with us a little longer.

Clockwise from top left: Nashold Twenty-Sided Barn, Columbia County; Adam Dunlap Farmstead, Dane County. Photo by Robert Dodsworth; Lutze Housebarn, Manitowoc County; Central Wisconsin State Fair Round Barn, Wood County. Photo by Mary Jane Hettinga; Zirbel Hildebrandt Farmstead, Dodge County. Photo by Carol Cartwright; Lockwood Barn, Dane County. Photo by James Steiner; Louis C. and Augusta Kriesel Farmstead, St. Croix County. Photo by Mead and Hunt, Inc.; Kliese Housebarn, Dodge County. Photo by Jim Kliese; Bedrud-Olson Farmstead, Dane County. Photo by Della Rucker

# Appendix I

## *Notable Wisconsin Barns on the National Register of Historic Places*

The following is a selection of unique and significant Wisconsin barns and farmsteads listed on the National Register of Historic Places (NRHP). (Nearly all of these properties are also listed on Wisconsin's State Register of Historic Places.) Following the name and location is the date the property was added to the register.

These barns were standing at the time of publication, but barns are disappearing at an alarming rate, and readers looking for them may see only a stone foundation or a lonely concrete silo. NRHP listing is primarily an honorary recognition and does not guarantee preservation or maintenance.

Barns and other notable historic buildings are typically nominated for the register by individuals or organizations (usually the owners) and not by the Wisconsin Historical Society. Readers who own a unique barn or intact farmstead that they believe is worthy of inclusion on the list should contact the Wisconsin Historical Society Office of Historic Preservation (see page 177).

Remember that most of these barns and farmsteads are on private property. Please respect the privacy of the owners.

### Columbia County

Holsten Family Farmstead. W1391 Weiner Road, Town of Columbus. Added to the National Register: 1992.

Nashold Twenty-Sided Barn. County Highway Z, .4 miles east of State Highway 146, Town of Fountain Prairie. Added to the National Register: 1988.

### Dane County

Bedrud-Olson Farmstead. 996 E. Church Road, Town of Christiana. Added to the National Register: 1999.

John Sweet Donald Farmstead. 1972 State Highway 92, Town of Springdale. Added to the National Register: 1984.

Adam Dunlap Farmstead. 9646 Dunlap Hollow Road, Town of Mazomanie. Added to the National Register: 2001.

Nicholas Haight Farmstead. 4926 Lacy Road, Fitchburg. Added to the National Register: 1993.

Lockwood Barn. Old Stage Road, Rutland. Added to the National Register: 1982.

Eric and Jerome Skindrud Farm. 3070 Town Hall Road, Town of Springdale. Added to the National Register: 1994.

University of Wisconsin Dairy Barn. 1915 Linden Drive, Madison. Added to the National Register: 2005.

**Dodge County**

Willard Greenfield Farmstead. N-7436 State Highway 26, Town of Burnett. Added to the National Register: 1992.

Kliese Housebarn. N366 County Road EM, Town of Emmet. Added to the National Register: 2008.

Zirbel-Hildebrandt Farmstead. W1328–1330 Highway 33, Town of Herman. Added to the National Register: 2007.

**Door County**

August Draize Farmstead. 814 Tru-Way Road, Town of Union. Added to the National Register: 2004.

Joseph Monfils Farmstead. 1463 Dump Road, Town of Brussels. Added to the National Register: 2004.

**Forest County**

Camp Five Farmstead. 5466 Connor Farm Road, Town of Laona. Added to the National Register: 1996.

**Green County**

Freitag Homestead. N7053 State Highway 69/39, Town of Washington. Added to the National Register: 2005.

Hefty-Blum Farmstead. W6303 Hefty Road, Town of Washington. Added to the National Register: 2000.

## Iowa County

Spensley Farm. 1126 Highway QQ, east of Highway 39, Town of Mineral Point. Added to the National Register: 1997.

Thomas Stone Barn. 7777 State Road 18–151, Town of Brigham. Added to the National Register: 2001.

## Iron County

Annala Round Barn. South of Hurley, Town of Oma. Added to the National Register: 1979.

## Jefferson County

Hoard's Dairyman Farm. North of Fort Atkinson, Town of Jefferson. Added to the National Register: 1978.

## Juneau County

William and Mary Shelton Farmstead. N2397 County Highway K, Town of Seven Mile Creek. Added to the National Register: 2004.

## Kewaunee County

George Halada Farmstead. E–1113 County Trunk Highway F, Town of Montpelier. Added to the National Register: 1993.

## La Crosse County

Chambers-Markle Farmstead. Junction of Trunk Highway 60 and County Trunk Highway X, Town of Richwood. Added to the National Register: 1991.

Carl August Mundstock Farm. U.S. 14/61, north side, east of State Highway 35, Town of Shelby. Added to the National Register: 1994.

## Manitowoc County

Lutze Housebarn. 13634 South Union Road, Town of Centerville. Added to the National Register: 1984.

## Oconto County

Daniel E. Krause Stone Barn. Northeast corner of County Trunk Highway S and Schwartz Road, Chase. Added to the National Register: 2000.

## Ozaukee County

William F. Jahn Farmstead. 12112–12116 North Wauwatosa Road, Mequon. Added to the National Register: 2000.

O'Brien-Peuschel Farmstead. 12510 North Wauwatosa Road, Mequon. Added to the National Register: 2000.

## Portage County

Severance-Pipe Farmstead. Pipe Road, one-eighth mile east of County Highway T, Town of Lanark. Added to the National Register: 1993.

## Richland County

John Coumbe Farmstead. Junction of Trunk Highway 60 and County Trunk Highway X, Town of Richwood. Added to the National Register: 1992.

## Rock County

Gempeler Round Barn. Southwest of Orfordville, Town of Spring Valley. Added to the National Register: 1979.

Gilley-Tofsland Octagonal Barn. Northwest of Edgerton, Town of Porter. Added to the National Register: 1979.

Risum Round Barn. Southwest of Orfordville, Town of Spring Valley. Added to the National Register: 1979.

## St. Croix County

Louis C. and Augusta Kriesel Farmstead. 132 State Highway 35/64, Town of Saint Joseph. Added to the National Register: 2009.

## Vernon County

George Apfel Round Barn. 11314 County Highway P, Town of Clinton. Added to the National Register: 2006.

## Walworth County

Anson Warner Farmstead. N9334 Warner Road, Town of Whitewater. Added to the National Register: 1998.

## Washington County

Messer-Mayer Mill. 4399 Pleasant Hill Road, Town of Richfield. Added to the National
Register: 2007.

## Waukesha County

Weston's Antique Apple Orchard. 19766 West National Road, New Berlin. Added to the
National Register: 1996.

Michael Wick Farmhouse and Barn. N72 W13449 Good Hope Road, Menomonee Falls. Added
to the National Register: 1988.

## Wood County

Central Wisconsin State Fair Round Barn. Junction of Vine Avenue and East 17 Street,
Marshfield. Added to the National Register: 1997.

# Appendix II

## *Map of Barn Locations*

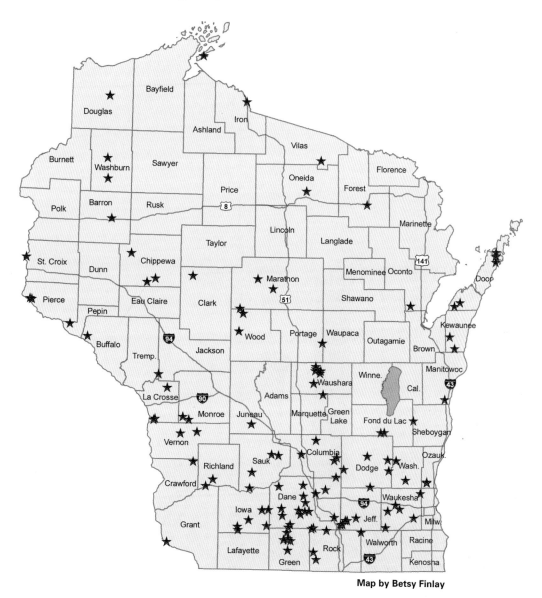

**Map by Betsy Finlay**

This map indicates the locations of the barns shown in the contemporary photos throughout this book and of those listed in Appendix I: Notable Wisconsin Barns on the National Register of Historic Places (page 185).

# Notes

## Chapter 1

1. U.S. Bureau of the Census, *Historical Statistics of the United States, 1789–1945.* Washington, DC, 1949.
2. *Wisconsin Blue Book* (Madison: Department of Administration, State of Wisconsin, 2007–2008), 615.
3. *Number of Farms and Land in Farms, Wisconsin and the United States* (Madison: Wisconsin Agricultural Statistics Service); *Dairy Statistics, Dairy Farms* (Wisconsin Agricultural Statistics Service, May 1, 2003).
4. *Number of Licensed Dairy Herds in Wisconsin* (Madison: Wisconsin Agricultural Statistics Service, October 1, 2009).

## Chapter 3

1. Eric Arthur and Dudley Witney, *The Barn* (Ontario: M. F. Feheley Arts Co., 1972), 85.
2. Eric Sloane, *An Age of Barns* (New York: Funk & Wagnalls, 1967), 42.
3. Richard W. E. Perrin, *Historic Wisconsin Buildings: A Survey of Pioneer Architecture, 1835–1870* (Milwaukee: Milwaukee Public Museum, 1962), 43.
4. J. H. Sanders, *Practical Hints about Barn Building* (Chicago: J. H. Sanders Publishing Co., 1893), 40.

## Chapter 4

1. O. S. Fowler, *A Home for All, or, The Gravel Wall and Octagon Mode of Building* (New York: Fowler and Wells, Publishers, 1854), 82.
2. Ibid., 174–178.
3. Ibid.
4. Richard W. E. Perrin, *Historic Wisconsin Buildings: A Survey of Pioneer Architecture, 1835–1870* (Milwaukee: Milwaukee Public Museum, 1962), 43.
5. Eric Sloane, *An Age of Barns* (New York: Funk and Wagnalls, 1967), 56–57.
6. Wilbur J. Fraser, *The Round Barn.* University of Illinois Agricultural Experiment Circular No. 230 (Urbana, IL: 1918), 13–15.
7. J. H. Sanders, *Practical Hints about Barn Building* (Chicago: J. H. Sanders Publishing Co., 1893), 100–101.
8. Ibid., 106–107.
9. Ibid., 108.
10. City of Marshfield, "Historic Preservation/National Register," http://ci.marshfield.wi.us/historic/national/roundbarn.htm.

## Chapter 5

1. Eric Sloane, *An Age of Barns* (New York: Funk & Wagnalls, 1967), 41.
2. H. J. Barre and L. L. Sammet, *Farm Structures* (New York: John Wiley & Sons, 1950), 225.
3. Lowell J. Soike, "Within the Reach of All: Midwest Barns Perfected," in *Barns of the Midwest*, ed. Allen G. Noble and Hubert G. H. Wilhem, (Athens: Ohio University Press, 1995) 160–165.

## Chapter 6

1. Fred L. Holmes, *Old World Wisconsin: Around Europe in the Badger State* (Eau Claire, WI: E. M. Hale and Co., 1944), 163.
2. Sears, Roebuck Catalogue (Chicago, 1908), 576.
3. W. A. Titus, "The First Concrete Building in the United States," *Wisconsin Magazine of History* 24 (1941): 183–188.
4. K. J. T. Ekblaw, *Farm Concrete* (New York: Macmillan Co., 1917), 1.
5. Ibid., 2.
6. Lowell J. Soike, "Within the Reach of All: Midwest Barns Perfected," in *Barns of the Midwest*, ed. Allen G. Noble and Hubert G. H. Wilhem, (Athens: Ohio University Press, 1995) 147–157; H. J. Barre and L. L. Sammet, *Farm Structures* (New York: John Wiley & Sons, 1950), 87–101.

## Chapter 8

1. Patricia Mullen, quoted in Eric Arthur and Dudley Witney, *The Barn* (Ontario: M. F. Feheley Arts Co., 1972), 180.
2. F. H. King, *Ventilation for Dwellings, Rural Schools and Stables* (Madison: published by the author, 1908), iii.

## Chapter 9

1. Bud (John) Carroll, personal correspondence, June 23, 2008.
2. Anne Bachner and David Lowe, "Application form, National Register of Historic Places. Thomas Stone Barn." United States Department of the Interior, National Park Service, May 15, 2000.
3. For additional information see Holly E. Smith, Susan Haswell, and Arnold R. Alanen, "Architecture and Science Associated with the Dairy Barn at the University of Wisconsin–Madison," Department of Landscape Architecture, University of Wisconsin–Madison, 2000.
4. William A. Henry, *Fifteenth Annual Report of the Agricultural Experiment Station of the University of Wisconsin* (Madison: 1898), 3.
5. Smith, Haswell, and Alanen, "Architecture and Science," 13.
6. Wisconsin National Register of Historic Places, www.wisconsinhistory.org/hp/register, accessed June 23, 2008.
7. Email correspondence with Voegeli Farm, July 14, 2008.

## Chapter 10

1. N. S. Fish, "The History of the Silo in Wisconsin," *Wisconsin Magazine of History* 8, no. 2 (1924–1925): 2.
2. *Oxford English Dictionary* (Oxford University Press, 1981).
3. Fish, "The History of the Silo in Wisconsin," 2.
4. N. S. Fish, "Farm Engineer Depicts Fifty Years of Steady Progress in Silo Building," first installment, mimeographed (Madison: n.d.), 3. (Fish was on the staff of the Agricultural Engineering Department, College of Agriculture, University of Wisconsin, from 1920 to 1925.)

5. Fish, "Farm Engineer," second installment, 5.

6. Fish, "Farm Engineer," third installment, 4.

7. Fish, "Farm Engineer," second installment, 3–4.

8. Fish, "The History of the Silo in Wisconsin," 3.

9. Ibid., 4–5.

10. "Beginning Our 100th Year," *Farm Journal,* 100, no. 4 (March 1976): 28.

11. Byron D. Halsted, *Barn Plans and Outbuildings* (New York: Orange Judd Co., 1893), 219–221.

12. Peggy Lee Beedle, "Silos: An Agricultural Success Story," *Giving Old Barns New Life* (Madison: University of Wisconsin–Extension, 2001), 6.

13. Jerry Apps, *The People Came First: A History of Wisconsin Cooperative Extension* (Madison: University of Wisconsin–Extension, 2003), 35.

14. Fish, "The History of the Silo in Wisconsin." *Wisconsin Magazine of History* 8, no. 2 (1924–1925): 9.

15. Robert Suter, *The Courage to Change* (Danville, IL: Interstate Printers and Publishers, 1964); Peggy Lee Beedle, "Silos: An Agricultural Success Story," *Giving Old Barns New Life* (Madison: University of Wisconsin–Extension, 2001), 14.

16. Funding Universe, "A. O. Smith Corporation," www.fundinguniverse.com/company-histories/AO-Smith-Corporation-Company-History.html.

## Chapter 11

1. Dale M. Lewison, "The Economic Geography of the Wisconsin Tobacco Industry" (master's thesis, University of Oklahoma, 1967), 50.

2. J. H. Sanders, *Practical Hints about Barn Building* (Chicago: J. H. Sanders Publishing Co., 1893), 11.

3. Chesla C. Sherlock, *Modern Farmyard Buildings* (Des Moines, IA: Homestead Co., 1922), 31.

4. Ibid., 32.

5. Gary A. Payne, "Interpretative Proposal: North German Farm Exhibit, OWW" (unpublished paper, Old World Wisconsin, Eagle, Wisconsin, Feb. 13, 1975).

6. Eric Sloane, *An Age of Barns* (New York: Funk & Wagnalls, 1967), 48–49.

## Chapter 12

1. "2008 State Agriculture Overview" (Madison: Wisconsin Agricultural Statistics Service, January 2009).

2. M. W. Mayer and D. W. Kammel, "2008 Wisconsin Dairy Modernization Survey" (Madison: University of Wisconsin–Extension, 2009).

3. Wisconsin Agricultural Statistics Service. *Changes in Wisconsin Dairy Industry, 1960, 1993, 1994* (Madison: National Agricultural Statistics Service, Wisconsin Department of Agriculture, Trade and Consumer Protection, 1995); "Dairy Statistics" (Madison: Wisconsin Agricultural Statistics Service, May 1, 2008).

4. "Dairy Statistics."

5. Ruth Olson, personal correspondence, July 11, 2008.

6. Adapted from Mary Humstone, *Barn Again! A Guide to Rehabilitation of Older Farm Buildings* (Des Moines, IA: Meredith Corporation and the National Trust for Historic Preservation, 1988), 8–9.

7. Charles Law, personal correspondence, July 29, 2008.

8. Quivey's Grove restaurant, "History," www.quiveysgrove.com.

9. Maggie Peterman, "Rebuilding the Barn," *Wisconsin Builder,* March 2009.

# Bibliography and Further Reading

**Books**

Apps, Jerry (Illustrations by Julie Sutter-Blair). *The Wisconsin Traveler's Companion.* Madison: Trails Books, 1997.

Arthur, Eric, and Dudley Witney. *The Barn.* Ontario: M. F. Feheley Arts Co., 1972.

Barre, H. J. and L. L. Sammet. *Farm Structures.* New York: John Wiley & Sons, 1950.

Billington, Ray Allen. *Westward Expansion: A History of the American Frontier.* New York: Macmillan Co., 1967.

Bogue, Allen G. *From Prairie to Corn Belt: Farming on the Illinois and Iowa Prairies in the Nineteenth Century.* Chicago: University of Chicago Press, 1963.

Clark, James I. *Wisconsin Agriculture: The Rise of the Dairy Cow.* Madison: State Historical Society of Wisconsin, 1956.

Ekblaw, K. J. T. *Farm Concrete.* New York: Macmillan Co., 1917.

Fowler, O. S. *A Home for All, or, The Gravel Wall and Octagon Mode of Building.* New York: Fowler and Wells, Publishers, 1854.

Halsted, Byron D. *Barn Plans and Outbuildings.* New York: Orange Judd Co., 1893.

Harris, Bill. *Barns of America.* New York: Crescent Books, 1991.

Holmes, Fred. L. *Old World Wisconsin: Around Europe in the Badger State.* Eau Claire, WI: E. M. Hale and Co., 1944.

Humstone, Mary. *Barn Again! A Guide to Rehabilitation of Older Farm Buildings.* Des Moines, IA: Meredith Corporation and the National Trust for Historic Preservation, 1988.

Hunter, Rebecca, and Dale Wolicki. *The Book of Barns: Sears, Roebuck and Co., Chicago* (reprint of 1919 catalog). Elgin, IL: Rebecca Hunter, 2005.

Johnson, Dexter W. *Using Old Farm Buildings.* Washington, DC: National Trust for Historic Preservation. National Park Service, Information Series, No. 46., 1989.

King, E. H. *Ventilation for Dwellings, Rural Schools and Stables.* Madison: published by the author, 1908.

Klamkin, Charles. *The Barn.* New York: Hawthorn Books, 1973.

Lampard, Eric E. *The Rise of the Dairy Industry in Wisconsin.* Madison: State Historical Society of Wisconsin, 1963.

Luxford, R. E., and George W. Thayer. *Wood Handbook.* Washington, DC: U.S. Department of Agriculture, 1935.

Martin, Lawrence. *The Physical Geography of Wisconsin.* Madison: University of Wisconsin Press, 1965.

Nesbit, Robert C. *Wisconsin: A History.* Madison: University of Wisconsin Press, 1973.

Noble, Allen G., and Hubert G. H. Wilhelm. *Barns of the Midwest.* Athens: Ohio University Press, 1995.

Perrin, Richard W. E. *The Architecture of Wisconsin.* Madison: State Historical Society of Wisconsin, 1967.

————. *Historic Wisconsin Buildings: A Survey of Pioneer Architecture, 1835–1870.* Milwaukee: Milwaukee Public Museum, 1962.

Radford, William A. (ed.) *Radford's Combined House and Barn Plan Book.* Chicago: Radford Architectural Co., 1908.

Rawson, Richard. *Old Barn Plans.* New York: Mayflower Books, 1979.

Sanders, J. H. *Practical Hints about Barn Building.* Chicago: J. H. Sanders Publishing Co., 1893.

Schafer, Joseph. *A History of Agriculture in Wisconsin.* Madison: State Historical Society of Wisconsin, 1922.

Schuler, Stanley. *American Barns, In a Class by Themselves.* Exton, PA: Schiffer Publishing, 1984.

Schumm-Burgess, Nancy and Ernest J. Schweit, *Wisconsin Barns.* Helena, MT: Farcountry Press, 2009.

*Sears, Roebuck Catalogue.* Chicago: 1908.

Sherlock, Chesla. *Modern Farmyard Buildings.* Des Moines, IA: Homestead Co., 1922.

Sloane, Eric. *An Age of Barns.* New York: Funk & Wagnalls, 1967.

————. *Our Vanishing Landscape.* New York: Funk & Wagnalls, 1955.

————. *A Reverence for Wood.* New York: Funk & Wagnalls, 1965.

Smith, Alice E. *The History of Wisconsin, Vol. I, From Exploration to Statehood,* William Fletcher Thompson (ed.). Madison: State Historical Society of Wisconsin, 1973.

Wysocky, Ken. *This Old Barn.* Greendale, WI: Reiman Publications, 1996.

## Monographs and Bulletins

Apps, Jerry. *Ethnic History and Beauty of Old Barns,* Giving Old Barns New Life series, Volume 1. University of Wisconsin-Extension, 1996.

————. *Wisconsin's Changing Farmsteads,* Giving Old Barns New Life series, Volume 2. University of Wisconsin–Extension, 1996.

Beedle, Peggy Lee. *Silos: An Agricultural Success Story,* Giving Old Barns New Life series, Volume 4. University of Wisconsin-Extension, 2001.

Beedle, Peggy Lee, and Geoffrey M. Gyrisco. *Barns and Barn Preservation: A Biography,* Giving Old Barns New Life series, Volume 3. University of Wisconsin-Extension, 1999.

Fraser, Wilber J. *Economy of the Round Dairy Barn.* Bulletin No. 143. University of Illinois, Agricultural Experiment Station, 1910.

## Articles and Periodicals

Apps, Jerry. "Rural Relics." *Wisconsin Trails* 41, no. 5 (October 2000): 50. "Beginning Our 100th Year." *Farm Journal* 100, no. 4 (March 1976): 28.

Doherty, Jim. "A Barn Is More than a Building: It Is a Shrine to Our Agrarian Past." *Smithsonian* 20, no. 14 (August 1989): 30.

Perrin, Richard W. E. "Wisconsin's Stovewood Architecture." *Wisconsin Academy Review* 20, no. 3 (Summer 1974): 2–8.

Titus, W. A. "The First Concrete Building in the United States." *Wisconsin Magazine of History* 24 (1941): 183–188.

## Reports

*Fifteenth Annual Report of the Agricultural Experiment Station of the University of Wisconsin.* Madison: 1898.

Fraser, Wilber J. *The Round Barn.* University of Illinois Agricultural Experiment Station Circular No. 230. Urbana: 1918.

Koegel, Richard G., and H. D. Bruhn. "Inherent Causes of Spontaneous Ignition in Silos," *Transactions of ASAE* [American Society of Agricultural Engineers] 14, no. 2 (1971).

Merriam, William R. *Census Reports, Volume V, Part I, Farms Livestock, and Animal Products.* Washington, DC: U.S. Census Office, 1902.

*Wisconsin's Changing Population.* University of Wisconsin, Bulletin No. 1642, Series 2426. Madison: 1942.

## Web Sites

www.jerryapps.com
Description of author's books, links to barn videos, interviews, and more

www.steveapps.com
More barn and farm photos

www.barnalliance.org
Information about an organization of barn preservationists

www.wisconsinhistory.org/hp
Wisconsin Historical Society, with detailed information about historic preservation, including how to list historic buildings on state and national registers

www.dalejtravis.com
List of Wisconsin round and multisided barns, with photos and locations

www.greencountybarnquilts.com
Green County, Wisconsin, barn quilt project, with information and locations of barns with quilt designs

www.quiveysgrove.com
History of Quivey's Grove Restaurant, which includes a stone stable

www.uwex.edu/lgc/barns
Information about Wisconsin's barn preservation program including lists of contractors, information about workshops, and bibliography

www.agriculture.com/barnagain
Detailed information about the BARN AGAIN! program

www.preservationnation.org
National Trust for Historic Preservation

www.oldworldwisconsin.wisconsinhistory.org
Site of several historic ethnic barns, farmsteads, and other rural buildings in a natural rural setting, located near Eagle, Wisconsin

www.stonefield.wisconsinhistory.org
Stonefield historic site, which includes a farmstead, old barns, silo, and Wisconsin's finest collection of historic farm equipment, located near Cassville, Wisconsin

www.townofchase.org
History and updates on the restoration of one of the country's last remaining all-fieldstone barns

# Index

sheep raising, 41
Shelton Farmstead, 187
shingles, *see* cedar shingles
Shivers, Alga, 60
Shivers, Thomas Level Ethridge, 60
silos: brick silos, *119*, *139*; construction of, 130–132, *131*, 138–141, *141*; emptying of, 4, 142, 144; Field Guide To, 145; fieldstone, *79*, *138*; filling of, 129–130, *129*, 136–137, *136*; metal silos, 141, *142–143*, 144–145; plastic bag silos, *131*; round silos, 137–141; spoilage of silage, 133, 137; square silos, 136–137; tile, *68*, *114*, 116, *128*, *129*, 140, *141*, *145*; trench silos, 132–136; University of Wisconsin–Madison dairy barn, *112*, *119*, 121; wood stave, 130, *131*, 139, *139*
Skindrud Farm, 186
Sloane, Eric, 55–56, 65, 104, 163
Smeaton, John, 82
Smith, Hiram, 44
smokehouses, 150–151
snug Dutch–roof barns, 67
social contact and farmstead arrangement, 159
solid-log construction, 22
sorghum, 41
sounds of barns, 101
Spensley Farm, 187
Spring Green round barn, 178, *179*
stables, *34*, *37*
State Fair Round Barn, *184*, 189
Steele, John, 134–135
stone barns, *see* fieldstone: barns; limestone barns
Stonefield historic site, 62, *62*, *74*
stonemasonry, *78*, 80, 89–91, *89*
stony soil, 79
stovewood barns, 26–27, *27*, *81*
straight lines in farming, 61, 138
Strang, Allen, xiii
surveying of land, 61
Swiss barns, 35–36, *35*, *36*
tamarack poles, 77
Thomas stone barn, 117–118, *117*, 187
Thomas, Walter, 118
three-bay barns (English barns), 23–24, *24*
threshing floor, 9, 33–35, 45, *46–47*, 48

tile silos, *68*, *114*, 116, *128*, *129*, 140, *141*, *145*
timber frame barns: broadax-hewn timbers, 75–76; erecting of, *90*, 94–97, *95*, *96*; fieldstone walls and, 78–80, *78*, *79*, *88–90*, 89–94, *94*; light construction, 83, *83*, 86–87; post-and-beam construction, 75–83, *76–81*, *77*, *96*, 123, 127; sawmills and, 75–77
time invested in barns, 11
tobacco farming, 40, *40*, 152–153, *152*
tornadoes, 3
town barns, *see* village barns
trench silos, 132–136
two-story barns: bank barns, *32*, 33–38, *34–37*, *44*, 45; dairy farming and, 39, 45; non-bank barns, 45, *46–47*, 48

University of Illinois round barn research, *55*, 57
University of Wisconsin–Madison: Center for the Study of Upper Midwestern Cultures, 174; College of Agriculture, 44, 133–138, 144; dairy barn, *103*, *112*, 118–122, *119*, *121*, 186; Forest Products Lab, 66

ventilation: cupolas and, 12, 104–106, 121; hay chutes and, 5; haymow floors and, 95; large barns and, 38, 116; round barns and, 59; Voegeli barn, 123
Vernon County, *40*, 60, *60*, 188
*Vierkanthof* farmsteads, 162
village barns, 48–49, *49*
Voegeli barn, 122–123, *124–125*, 127
Voegeli, Jake, 123

Walworth County, 26, 188
Warner Farmstead, 188
Washburn County, *81*, *173*
Washington County, *136*, 189
Waukesha County: cedar-shingled horse barn, *72*; converted barn, *54*, 176, 178; fieldstone barns and silos, *79*, *89*, *138*; Finnish barns, 25, *25*, *149*; gable-roof barn, *63*; German barns, 28–29, *31*, 77, *96*; National Register barns and farmsteads, 189; Norwegian barns, 22, *24*, 150; Norwegian corncrib, *150*

# About the Author and Photographer

**Jerry Apps** is professor emeritus at the University of Wisconsin–Madison and the author of many books on rural history and country life. Jerry's nonfiction books include *Horse-Drawn Days, Old Farm, Every Farm Tells a Story, Living a Country Year, When Chores Were Done, Humor from the Country, Country Ways and Country Days,* and *Ringlingville USA.* His historical fiction includes *The Travels of Increase Joseph, In a Pickle,* and *Blue Shadows Farm.* He received the Council for Wisconsin Writers' 2007 Major Achievement Award and the Wisconsin Library Association's 2007 Notable Wisconsin Author Award.

Jerry was born and raised on a small farm in Waushara County, Wisconsin, where he spent countless hours working in the barn, milking and feeding cows, helping to store hay in the haymow, and appreciating the barn's importance to the life of a farm.

**Steve Apps** is an award-winning photojournalist with twenty-five years in the newspaper industry. As a *Wisconsin State Journal* staff photographer he has covered a wide range of assignments including the Green Bay Packers and University of Wisconsin–Madison sports. In 2008 he received the Pro Football Hall of Fame's prestigious Dave Boss Award of Excellence; his photo "First Down" was selected as Photograph of the Year for the 2007 season. In his off-time Steve loves to travel the state documenting Wisconsin and all its beauty, including farmsteads and barns.